E. Walter Maunder

The Indian Eclipse

The Breitish Astronomical Association

E. Walter Maunder

The Indian Eclipse
The Breitish Astronomical Association

ISBN/EAN: 9783741103520

Manufactured in Europe, USA, Canada, Australia, Japa

Cover: Foto ©ninafisch / pixelio.de

Manufactured and distributed by brebook publishing software
(www.brebook.com)

E. Walter Maunder

The Indian Eclipse

BRITISH ASTRONOMICAL ASSOCIATION

THE INDIAN ECLIPSE

1898

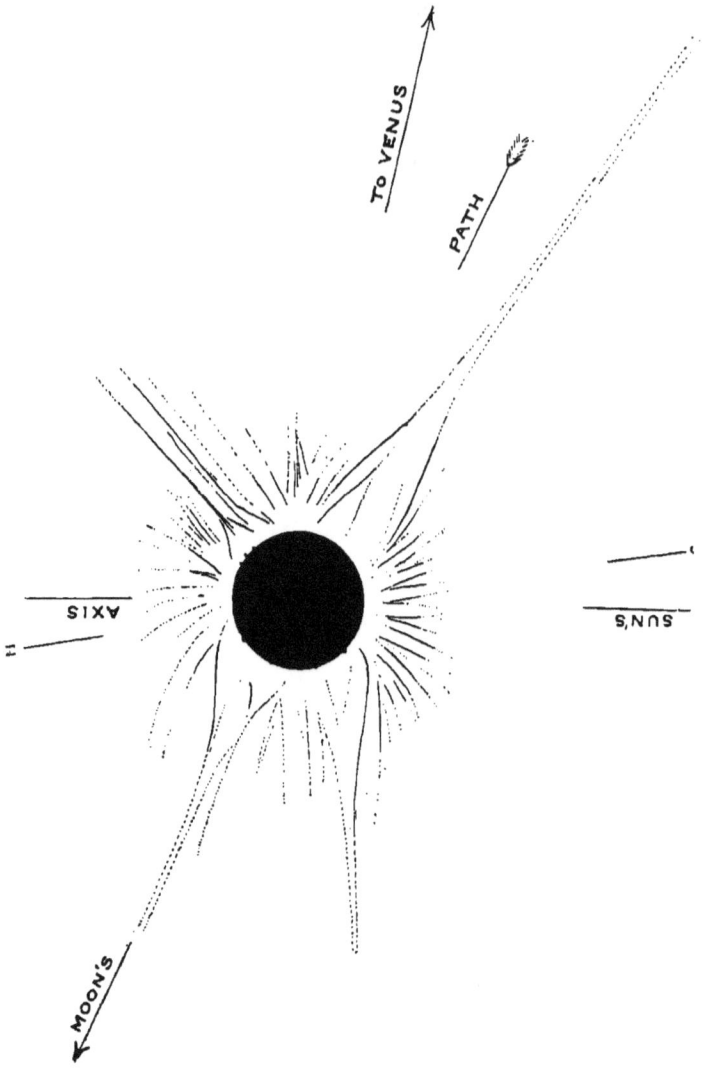

TO VENUS

PATH

AXIS

SUN'S

MOON'S

THE OUTLINES OF THE CORONA, FROM THE PHOTOGRAPHS.

Drawn by Mr. W. H. Wesley.

BRITISH ASTRONOMICAL ASSOCIATION

THE INDIAN ECLIPSE

1898

REPORT OF THE EXPEDITIONS ORGANIZED BY
THE BRITISH ASTRONOMICAL ASSOCIATION TO
OBSERVE THE TOTAL SOLAR ECLIPSE
OF 1898, JANUARY 22

EDITED BY

E. WALTER MAUNDER, F.R.A.S.

London

HAZELL, WATSON, AND VINEY, Ld.

1, CREED LANE, LUDGATE HILL, E.C.

1899

PRINTED BY
HAZELL, WATSON, AND VINEY, LD.,
LONDON AND AYLESBURY.

PREFACE

THE British Astronomical Association was founded in October 1890, with the following objects :—The association of observers, especially the possessors of small telescopes, for mutual help, and their organisation in the work of astronomical observation. This is effected by the " Observing Sections," under experienced Directors. The circulation of current astronomical information by a monthly journal and by the memoirs of the various Sections ; the encouragement of a popular interest in Astronomy by the monthly meetings of the Association, held not only in London, but also at the seats of the Branches —Manchester, Glasgow, Edinburgh, Sydney, and Melbourne. The Association now numbers considerably over eleven hundred members, and possesses an astronomical library and a large collection of astronomical lantern slides for loan to members ; and an observatory for education and research will shortly be available.

The following Report gives an account of the second expedition organised by the Association to observe an eclipse of the sun. Any profits resulting from its sale will be devoted to the aid of future similar expeditions, the next of which— namely, that to Portugal, Spain, Algeria, and the United States, to observe the eclipse of 1900, May 28—is now being arranged for.

The entrance fee to the Association is five shillings ; the annual subscription is half a guinea. All inquiries should be made to

THE ASSISTANT SECRETARY,

26, MARTIN'S LANE,

CANNON STREET, E.C.

1899, *June* 28.

CONTENTS.

LIST OF ILLUSTRATIONS.

xi

THE INDIAN ECLIPSE, 1898.

CHAPTER I.

INTRODUCTORY.

THE following report is the record of the second eclipse expedition organised by the British Astronomical Association. The first—that to Norway in 1896—was a new departure in astronomical expeditions. It was the first time that a company of anything approaching its magnitude had been organised upon the admirable Pickwickian principle,* and composed entirely of volunteers, for the purpose of observing a total eclipse. And though it was unfortunate in the weather conditions on the day of the eclipse, and consequently was without those scientific fruits which had been hoped from it, it was most successful in every particular which depended upon the members themselves. For its indirect results have been great. It resulted in a considerable increase in the membership of the Association; in greatly promoting the union of the members; and in the creation and cementing of many friendships which promise to be lifelong. It was a new effort in co-operation, and, thanks to the hearty goodwill and ready spirit of help that was shown on every hand, it was, so far as the conditions allowed. a most successful one. It impressed upon the young Association an increased sense of its own power for usefulness, and it brought it into prominence before the public.

The first thought of the Eclipse Committee, when it became necessary to set on foot preparations for the eclipse of January 1898, was, if possible, to organise an expedition on similar lines to those of the first. The pleasure which had been derived in the Norway expedition from travelling together in the same ship and with so many friends, bound on the same errand, and visiting together so many places of interest, suggested that

* "This Association cordially recognises the principle of every member of the Corresponding Society defraying his own travelling expenses." (*Posthumous Papers of the Pickwick Club*, Chapter I.)

1

a similar plan was worth trying in the case of India. After a multiplicity of negociations this plan had to be given up, and arrangements were made for the conveyance of observers by the P. and O. mail steamers. The secretary of the Committee was most courteously received by the manager and by the secretary of the P. and O. Company, and every concession in their power to give was readily made. In particular, the eclipse parties have to thank them for their arrangements with regard to the transport of their instruments, of which every possible care was taken, and for the way in which every effort was made to meet their convenience. The Committee would like to take this opportunity of expressing their indebtedness to them.

The next matter to be arranged for was the observing station in India. Here two or three difficulties had to be met. First, the line of totality passed through no large town until it reached the banks of the Ganges, and it would be therefore necessary to set up and provision a camp. For this Masur, a small village on the Southern Mahratta Railway, was chosen, firstly because the totality was longer there than at any other accessible station except that of Viziadrug, already occupied by Sir Norman Lockyer; and next because no other party had announced their intention of going there. The directors of the South Mahratta Railway, on learning our decision, most generously offered free passes over their lines for all members of our party, and free transport for our astronomical instruments.

The difficulty felt in the selection of an astronomical site in a country altogether unknown to us was overcome by the great generosity of two members of the Association—Miss Harriet and Miss Eliza Wigram. They presented the Committee with £125, a gift which enabled the Committee to meet the expenses of a pioneer, who should go out in advance of the other observers and select the precise place for their camp, and complete all other arrangements necessary on the spot. His experience of India suggested Mr. C. Thwaites as most suitable for this important duty, which he very kindly undertook to fulfil. In the meantime, camping arrangements were being most kindly negociated for us by Mr. Henry Cousens, Superintendent of the Archæological Survey of India, and up to within a fortnight of the time fixed for the starting of Mr. Thwaites it was supposed that everything was satisfactorily settled. Unfortunately the spread of plague in the Satara district, in which Masur is situated, had been so severe that we were strongly urged by the Bombay Government, through the medium of the India Office, to abandon our idea of going to Masur. We therefore telegraphed to Mr. Cousens to cancel the contract which he was about to make for our station in the Satara district.

This disappointment, occurring at the last moment, made it clear to us that we must abandon the idea of forming but a single camp. And it seemed best, from such information as we had at hand, that those observers who had the most complicated instruments should go out in R.M.S. *Ballaarat*, with Mr. Thwaites, and should form a camp at Talni or Pulgaon—two villages on the G.I.P. railway, where the latter cuts the line of totality—whilst the second party, coming out in R.M.S. *Egypt*, should try and occupy some position near the banks of the Ganges where they might obtain hotel accommodation in some

BUXAR RAILWAY STATION.

large town, which would entirely do away with the need of camping out at all. We therefore addressed ourselves to the India Office, asking them to recommend us to the different local governments in India for their assistance in the position in which we found ourselves placed; and we left England necessarily without having heard what preparations would be made for us, or precisely where we should have to take up our stations.

E. WALTER MAUNDER }
JOHN M. BACON   } *Secretaries.*

CHAPTER II.

THE EXPEDITION AT TALNI.

IT has been the subject of many discussions amongst us since our return as to why it was that the five † of us who made up the first of the two expeditions found ourselves on board R.M.S. *Ballaarat* at Tilbury on 1897, December 8th. It would have saved us a week in time, it would have cost us no more in cash, and, above all, we should have escaped a decidedly severe "bucketing in the Bay," had we taken the course preferred by many of our fellow-passengers, and gone by the overland route to Marseilles. No satisfactory answer has ever been given to the question, but at any rate we gained experience by it, for we encountered one of the most serious storms of the year in the dreaded Bay. But from passing Cape St. Vincent until we reached Bombay, on January 3rd, the weather was never rough enough to interfere with such astronomical observations as circumstances on shipboard permitted us to make.

Any description of the ordinary incidents of ship life, or of the places at which our vessel touched, would be out of place here, for it would not differ materially from that which might be given by any traveller along the same route. But we may be said to have begun our astronomical work when on December 12th we scrutinised the great group of sun-spots, then very near the centre of the disk. Day by day we made independent determinations of our latitude at noon, and were gratified to find that the errors of the ship's officers were inconsiderable. After leaving Marseilles on December 16th. we had the pleasure of recognising Mercury in the evening twilight, and for the next fortnight the search for him immediately after sundown was an unfailing source of interest. After leaving Brindisi, on December 19th, an entirely new object caught our attention. This was the Zodiacal Light. The difference between this broad glowing beam, seen so constantly evening after evening, and its pale uncertain counterfeit which on rare occasions we had seen in England, was most striking, and came quite like a new revelation to those of us who were making their first voyage into

---

* By E. Walter Maunder, *Secretary*.
† The five were—Mr. J. Evershed, Mr. and Mrs. Walter Maunder, Mr. J. P. G. Smith, and Mr. C. Thwaites. Captain P. B. Molesworth, R.E., joined us from Ceylon in the camp at Talni, completing our party.

4

the tropics. It soon became a regular item of the evening's programme to map out the contour of the Light and form estimates of its intensity; and we supplemented these by a few observations of the morning branch of the Light before sunrise. The voyage down the Red Sea, December 24th to 27th, was especially memorable. The early mornings afforded Mr. Evershed welcome opportunities for testing his slit spectroscope. His cabin was on the east, the port side, of the ship, and his berth pointed to the porthole. It was thus very simple for him to affix his spectroscope to the underside of Mr. Thwaites' berth, which was above his own, so that the light from the porthole fell directly on the slit. The trifling inconvenience which this otherwise excellent arrangement caused Mr. Thwaites was, of course, cheerfully borne in the interests of science, and Mr. Evershed obtained some very satisfactory photographs of the Fraünhofer lines as a test of the focus of his instrument. In the evening, just before sundown, the spectroscopes were brought into work on the starboard side, the great telluric bands coming out like storm-clouds in the spectrum of the setting sun; whilst the moment when the sun's image was diminished to a single narrow segment was most admirable for bringing my own little spectroscopic opera-glass into perfect focus. Then, as the sun's rim dipped, we crossed to the port side, and watched the sharply defined outline of the earth's shadow creep upward and obliterate the rose and purple of the eastern sky. Next came the search for Mercury, which was usually found by the naked eye within half an hour of sunset, and ere he sank the starry host began to gather. A little later, and the Zodiacal Light came out like a second twilight, soon to be followed by the less brilliant radiance of the Milky Way. The two met on the horizon, their axes probably intersecting at the place of the sun, now sunk considerably below it. Here the Light quite overpowered its rival. When full darkness had come on, the two were seen intersecting each other at almost the opposite point of the heavens; but here the Light had faded to the most evanescent faintness. In the early evening both objects cast a broad path of light on the waters. That from the zodiacal glow soon faded out, but as the night deepened the star-shine from a score of brilliants made miniature moon-tracks to every part of the horizon. We were now travelling nearly due south, and a mere glance at the sky on each successive evening was quite sufficient to show how rapidly the Pole Star sank, and the new stranger southern constellations rushed up from the under-world.

> "And the Southern Cross, like a standard flying,
> Hangs in the front of the tropic night;
> But the Great Bear sinks like a hero dying,
> And the Pole Star lowers its signal light."

We did not greet the Southern Cross till we had reached
" The Exile's Gate " (Bab-el-mandeb), but it was a morning
much to be remembered when we first saw the most celebrated
asterism of the nether sky, and with it in Sagittarius the most
brilliant regions of the galaxy. Most brilliant of all, queen of
the waning night, shone Venus, to be lost in a few minutes in
the rays of the rising sun, and not to be again seen until, as if
anxious to take his place, she showed herself on the eventful day
even before her lord the Sun had completely withdrawn himself
behind the curtains of eclipse.

The *Ballaarat* had her merits, but speed was not one of
them. Most of the vessels of the line would have got us in by
the Saturday; some would have even reached their destination
by the Friday; but we were sufficiently thankful to be landed
on the Monday morning, January 3rd. A number of letters
were at once delivered to us. Mr. F. S. Bullock, the Chief
Commissioner of the " Hyderabad Assigned Districts "—that is
to say, the province of Berar—informed us that he had been
asked to make arrangements for our reception at the little
village of Talni, and that he had ordered a standing camp of
tents to be got ready and furnished. As, however, he would be
on duty away from headquarters, he desired us to communicate
further with Captain Horsburgh, the Deputy Commissioner for
the Amraoti District, in which Talni is situated. A second letter
was from Captain Horsburgh, and gave us further information
as to the preparations that were being made for us. A third
was from Assistant Commissioner Mr. D. O. Morris, Lieutenant
R.A., and told us that he had been personally charged with our
reception, and desired that one of our number should come
down at once to Talni, select the precise site for our observatories,
and see that all was in order. All anxiety was therefore put to
rest, as it was evident that every possible pains was being taken
to help us, and to secure a suitable and comfortable camp.

Before we had even landed a gentleman previously unknown
to all our party came on board to greet us. Mr. Smith had
intended, whilst in India, to visit some friends at Hyderabad;
and the latter being unable to come down to Bombay to meet
him at landing, had desired a friend, Col. Goodier Adye, then
staying in that city, to represent them. Col. Adye, who we
found had the reputation of being one of the most experienced
men in India in the knowledge of the native languages and
ways, at once began to make himself useful. He took all the
troubles connected with our luggage out of our hands. He
gathered the numerous boxes together, passed them through
the customs, ordered about the coolies, and exercised a firm
control over the drivers of the bullock gharries, who seemed at
first disposed to be troublesome. He piloted Mr. Thwaites to
the Victoria Station, and assisted him in his negociations with

Mr. Evershed.   Captain Molesworth.

Mr. Maunder.   Mrs. Maunder.   Mr. Smith.   Mr. Thwaites.

AFTER THE ECLIPSE, TALNI. OBSERVERS GARLANDED BY THE NATIVES.

the G.I.P. We had already received a promise of help and liberal treatment before we left England, Mr. Thwaites having corresponded with the secretary on our behalf. But the actual arrangements took some hours to make. They resulted, however, in our having a special through carriage allotted for us down to Talni instead of having to change at Bhusawal, and a special goods truck was set apart for our instruments and heavier personal luggage, all of which were at once placed in the van and carefully locked up before our eyes, so that we were assured they would not be tampered with until they were actually delivered at our destination. In addition to these favours a very substantial reduction was made to us in the company's charges. We may add here that the same consideration was shown us on our return, a special truck being detailed for our instruments at Talni railway station when sending them back to Bombay.

Mr. Thwaites, having undertaken to be our forerunner at Talni, started by the evening train, after about as busy a day as he had ever experienced. The other four of us were to remain until summoned by a telegram from him, and in consequence did not leave until the evening of Wednesday, January 5th. In the meantime Col. Adye made himself our guide to Bombay and the neighbourhood. Now that we were actually in Bombay all fear or thought of the plague seemed to vanish. On Tuesday evening we drove through the native bazaar, and wondered, as we viewed the crowds, if this was Bombay "empty and desolate," what it could be like when in the full tide of health and prosperity. Col. Adye came to see us off on the Wednesday evening, and we took leave of him, he promising to join us in our camp the day before the eclipse. We never saw him again. He attempted to fulfil his promise, but was taken ill on the way down and was unable to complete the journey, and he died in the week following the eclipse. This sad event had been entirely unexpected by us, as he had written a long letter to Mr. Smith only the day before he started to come down to us, and it cast a great gloom over our last days in camp.

There is no need for us to enlarge on the wonderful railway line from Bombay up the Ghâts. We travelled at night, of course; but a brilliant moon, now almost full, lit up the wild mountain scenery like a northern day. The journey to Talni took nearly eighteen hours, so that we did not arrive till about three o'clock on Thursday afternoon. The country that we traversed by daylight was very different from that which the moon had shown us. We were crossing the great plain of Central India, flat, bare, monotonous, and above all dusty; and we noted with many forebodings that, though the sky was absolutely free from any trace of cloud, a bright dust-haze whitened the entire sky. After passing Badnera, the last great junction on our route, the country became a little better wooded, and possibly in consequence we noted that the dust-haze was less apparent and the sky was bluer.

Talni proved to be quite a small village of a very ordinary type ; but not entirely without some picturesqueness in appearance. A mass of small mud huts closely packed together, and intersected by the narrowest of lanes, save where a number of lanes converged to an open space in the centre of which was the village well. Two buildings of greater consequence were alone to be seen : the house of the *patel* or head man, neat, white and square, with battlemented roof; and the mud fort, the latter in utter ruin.

Of the smaller huts and their inhabitants the photograph gives a good impression. It was taken early one Sunday

A HUT IN TALNI VILLAGE.

morning by the senior member of our little party. That it was early may be readily inferred from the muffled-up appearance of some of the group. For though the thermometer ran up during the day to 90° in the shade, the nights and early mornings were most decidedly chilly, and the natives were evidently not proof against the cold. It struck us as interesting to note how they seemed to find it amply sufficient when the temperature was low to wrap up their heads and cover their mouths. The body and limbs might take care of themselves.

Talni village was some little distance from Talni station, though a few huts and houses in which the station officials lived stood close to the line. The station was of the smallest, and did not even possess a platform, and in the ordinary way only the slowest trains stopped there—two each way during the twenty-four hours. Still the station buildings were at least as large and substantial as would be found in a village of similar size in England.

Mr. Thwaites met us with a small army of coolies, most of whom, however, had to go away unburdened, as our instruments did not arrive till the following day. For us a couple of bullock carts, one of which is seen in the photograph, which was taken on a later visit to the station, were provided; and in them we rode up to the camp which had been prepared for us, Mr. Thwaites pointing out as we went that a road a mile and a quarter long had been cleared for us from the station to the camp.

At the camp we met our host, Mr. D. O. Morris, to whose thoughtfulness and energy we already felt we owed much, a sense which was to increase every day. The sight of the little

TALNI RAILWAY STATION.

camp itself was sufficient to show us that everything possible for our help and comfort would be done. For its site a space of ten or twelve acres of cotton and jowari fields had been cleared. The dwelling camp occupied the south of this area, where stood a pretty grove of tamarind and mango trees, under the tallest of which the mess tent had been pitched, whilst the sleeping tents ran in a straight line due east and west; so far as possible also under the shade of trees. Paths from tent to tent had been carefully prepared, and a path about one hundred and fifty yards in length and stretching due north brought us to the line of our observing huts, placed quite out in the open, and as far as possible from any tree which could spoil our horizon. Before each tent stood a lamp-post, so that we might run no risk of tripping over tent ropes or stumbling into ditches when returning from our little observatories at night. Within, our tents if not luxuriously were at least sufficiently well furnished, and for almost three weeks made us most pleasant and comfortable homes.

Our sleeping tents ran in a straight line, as near as might be due east and west. The tent farthest to the east was tenantless until a day or two before the eclipse, when Capt. Grant came to visit our host, Mr. Morris. Next was the tent reserved for Capt. Molesworth. Mr. Evershed's, Mr. Thwaites', and Mr. Smith's tents followed in the order given; then came a large square tent for our instruments, packing-cases, etc., and farthest to the west our own tent.

Our observing huts were also arranged in a straight line due east and west ; but they lay some 150 yards away to the north, and somewhat to the east of the dwelling camp, so that the latter in no way interfered with the view. Our huts were each about 12 feet square in area, and about 7 feet in height. A strong wooden framework was first constructed, and then this was filled in with bamboo matting, which could easily be taken down as required. Each hut was roofed by three long shutters made of bamboo matting covered with grass, which were lifted on or off in a moment or two by the coolies, and which when in place kept the huts surprisingly cool even in the hottest part of the day. Our telescopes were provided with brick piers, and the floors of our huts were especially substantial ; the ground being first levelled, and then a course of bricks laid down, over which a layer of cement was placed. This floor was regularly washed out every morning, so as to keep the interior of the hut as free from dust as possible, for dust we saw was likely to be by far our worst enemy.

Mr. Evershed's requirements were somewhat different from our own, but were met as fully as possible. It became necessary for him to have a pit dug in the rear of his hut, and here was found the only venomous snake that we saw during our stay in Talni. We found it somewhat of a disappointment, after the thrilling tales of tigers, panthers, and hamadryads, with which our friends from India had tried to cheer us, both before we left England and on our voyage, to come down to one poor little two-foot-long snake, and a brace or so of jackals, whose presence was only indicated by their howls. The snake was however worth noting, if only for the original piece of natural history which we gathered concerning it. Its tail is short and blunt, resembling its head in shape ; and the natives believe that it interchanges head and tail every six months, so that the end which is head in January is tail in July ! Whether the change is instantaneous or gradual was a point upon which I obtained no light.

Our observing huts—the plan of which Mr. Thwaites and Mr. Morris had worked out together—were four in number. Mr. Smith held the most easterly, our hut followed next, Mr. Thwaites came third, whilst Mr. Evershed chose his nearest the sunset.

It will be easily understood that the fifteen days which were at our disposal for the work of preparation were very busy ones, and even if Talni or its neighbourhood had had many attractions to offer to the sight-seer, we could not have taken much advantage of them. Our daily routine was much as follows :— Unless we had been observing very late the night before, we had "chota hazri" soon after daybreak, and rose shortly after. Until about 10 o'clock we worked in our observatories, when we returned to our tents for our baths, and to get ready for breakfast. Some of us found it convenient to change into lighter clothes at this opportunity, so rapid was the rise of temperature as the sun approached the meridian. Breakfast was about 11 or 11.30, and was a long leisurely meal, the day being now very hot, but the heat being relieved by the "devils," the local term for little miniature whirlwinds that sprang up about noon, each lasting but a few seconds, and travelling a few yards, but keeping the air from stagnation. From breakfast to tea-time was the lazy time of the day, the time for reading or writing, or even for a nap. At all events, so far as might be, we kept in our tents or under our trees, and if we worked, occupied ourselves with matters that could be carried on there. Afternoon tea was an informal and movable feast, but by half-past four we had generally all found our way to our meeting tent. From that time to sundown we were usually all hard at work in our observing huts. During the first hour after sundown the Zodiacal Light which had so impressed us on our voyage usually again claimed our attention, and we made several attempts, unfortunately unsuccessful, to photograph it and its spectrum.* Next, as the stars came out, we determined or tested the adjustments of our instruments. Then, when the darkness had become complete, my wife set about her task of photographing the Milky Way with our little camera, the field of which was some 36° square. Dinner time was about half-past eight, after which, if moonlight did not interfere, we renewed our photographic work until we had either given as long an exposure to a plate as we desired, or the field upon which we were engaged had moved to an inconvenient position, or the moon had begun to show herself.

Those still cold nights under the bright Indian stars will not soon be forgotten by us. The heavens claimed chief attention. It was not only that during our entire stay in camp, no single

---

* Capt. Molesworth subsequently succeeded in photographing it at Trincomalee.

wisp of cloud ever dimmed their glory by day or night. Nor was it only the appearance in the south of stranger constellations, attractive both by their brightness and their novelty. The groups that had been familiar to us in England wore, many of them, a new aspect. The thin evasive Gegenschein could be detected between Pollux, Procyon, and Præsepe; whilst here and there were hints and suggestions of broad sheets of faintest luminosity, evanescent outshoots and streamers from the galaxy, or independent structures, ghost-like in their elusive faintness, but yet continually asserting the reality of their presence. To go back to childish fancy, or to mediæval thought, which looked upon the unfathomed sky as the inner side of a vast vault, the upper and outer side of which was the floor of heaven, we might have expressed what we saw by saying that " the floor of heaven was wearing thin."

If the heavens were brilliant, the earth was dark enough, though a few lights broke the gloom. Close at hand an occasional flash marked when a ray from the observer's lantern fell on the drawn sword of our police guard as he paced to and fro before our little observatories. In the distance burned the fires by which the coolies stretched themselves to sleep ; and farther to the south we could see one or two of the street lamps of our little canvas village. And for the most part it was intensely still. The undercurrent of sound, which was hardly ever lost in our night watches at Greenwich, was not detected here ; and such sounds as came—the official cough of the "chowkidar," which greatly excited our sympathy until we learned its significance, the howl of a wandering jackal, and the much execrated lamentations of the pariah dog from the distant village—were each distinct, clear, and sharp.

Such was the general current of our lives during the fortnight before the eclipse. There was much to do, and we had no time to feel bored or dull. The incidents which diversified our life were few, and can soon be recounted.

Mr. Thwaites, our pioneer, reached Talni on Tuesday, January 4th. The rest of us followed on Thursday, January 6th. The next day our instruments arrived, and the work of unpacking was at once commenced. Early the following morning a small partial eclipse of the moon was due, and we observed it to the best of our ability, my wife securing a series of photographs of it with her little camera, which served to test its focus. As none of the mountings were up by this time, it was simply fixed to a board and tilted up towards the moon by means of a chair.

A week later the only accident which befel any of our party happened to Mr. Thwaites, who, eagerly chasing a pretty lizard, was struck in the face by a falling bamboo. Fortunately the frontal bone received the chief weight of the blow, which was a very severe one : a fraction of an inch lower down, and the

THE DWELLING CAMP AT TALNI.

eye would inevitably have been destroyed. As it was, no permanent injury resulted, and Mr. Thwaites was sufficiently recovered to carry out his full programme at the eclipse.

The following Sunday, January 16th, saw our party completed by the arrival from Ceylon of Captain P. B. Molesworth, R.E., who had most kindly promised to help us in any way that he could in our work on the eclipse, and who redeemed his promise to the full.

The next morning Mr. Evershed, Captain Molesworth, and ourselves paid a visit to our "next-door neighbours," Captain Hills and Mr. Newall, who were encamped at Pulgaon, six miles farther along the line, and just across the frontier of the Central Provinces. We went by rail, but not by train, as there was none which could possibly serve us. So a couple of trolleys were put at our disposal, each of which was propelled by a couple of coolies, who ran along the actual metals at a speed of about seven miles an hour without ever missing their footing. This mode of progression was new to us all, and we found it the pleasantest that we had experienced in India, except perhaps whilst we were running across the bridge over the river Wardha ; for, as the bridge was not planked between the rails, we looked down from our seat on the front of the trolley and saw a clear drop before us of nearly one hundred feet to the bed of the river below.

At Pulgaon we received a very hospitable welcome from the official party, and were shown their instruments and other preparations. Amongst these I think we most admired the ingenious mounting which Mr. Newall had devised for his double-slit spectroscope, in which the collimator was arranged so as to be parallel to the polar axis. We envied our friends also their possession of a wooden dark-room—clean, roomy, comfortable, light and light-tight. Our own dark-room at Talni was of the quaintest form and appearance. It was a structure of wattle and daub, and its dun walls bowed and bent in all kinds of fantastic curves. The photograph on p. 18 shows the great earthen vessel which formed our water cistern, and the brick platform that the coolie mounted in order to fill it. Still, though queer-looking and unsymmetrical, the little hut served our turn for such photography as we were obliged to carry on during the day. At night we found our tents more comfortable for developing or plate-changing.

One other feature of the Pulgaon camp caught our fancy. This was the pretty flag made by Mrs. Hills, which floated over their chief tent, and bore " a corona argent, with prominences gules proper, round a sun effaced sable, in a field azure." Otherwise we were obliged to confess, when we returned to Talni, and our host, Mr. Morris, anxiously questioned us as to whether there was anything in the Pulgaon camp which we lacked, that, comfortable and well-appointed as everything was there, there was nothing in which better provision had not been made for our comfort than for theirs. We mentioned the flag as the one solitary point in which our camp was the inferior; and on the next day but one our standard, of fully twice the size of that at Pulgaon, floated proudly from the tree above our mess tent, and gave a larger corona to the breeze. Besides the flag we could think of nothing whatsoever but the white stones that marked out the paths from tent to tent at Pulgaon. This was a refinement which we really did not need in our camp, as our paths were carefully cut; but the omission, if such it could be called, was at once repaired, and long lines of white stones marked out all our ways before a second sunset.

DEVELOPING HUT, TALNI CAMP.

Later on in the week we received a visit from Mr. Elrington, the Superintendent of Telegraphs at Nagpur, who gave us a most material piece of assistance by arranging that we should have a special telegraph station at Talni for the day of the eclipse and for a day or two before and after. Otherwise it would have been necessary for one at least of our party to have taken the afternoon train to Wardha, twenty-five miles down the line, and to have returned about midnight. On Friday this temporary telegraph office was opened for the first time, and we received time signals from Madras at four o'clock in the afternoon. As we drove down to the railway station we met on the road, usually so completely deserted, an endless procession of natives; many of them the servants of the Resident from

Hyderabad, Sir Trevor Chichele-Plowden, who was on his way to Talni; the others, sight-seers from many miles round.

Mr. Bullock, the Chief Commissioner of Berar, for whose kindness to our party we would here wish to express our thanks and great indebtedness, had already arrived; together with Captain Horsburgh, the Commissioner of the Amraoti District; and long lines of tents were rising on the southern edge of our camp area. To those of us who were new to India, the sight, familiar enough to residents, of the long trains of camels and of the speed and skill with which numerous tents were pitched and furnished, and the suddenness with which all disappeared after the eclipse was over, was one full of interest.

CRESCENT SUNS ON THE PAVEMENT DURING THE PARTIAL ECLIPSE.

The eventful morning came at last, and a sky without the slightest ghost of a cloud relieved our minds of every possible anxiety as to interruption from weather. We were spared all fear of a repetition of the ill fortune of Norway. Our thoughtful host fixed the breakfast hour at 10, that we might have plenty of time to begin our observations with the first contact. The partial phase had, however, very little interest for us, Mr. Smith alone intending to observe the times of contact. The time of waiting till totality came on was long and trying, and we occupied it as far as we could by continued rehearsals of our programme. At 11.45 Mr. Smith warned us that the eclipse had commenced. At this Mr. Morris distributed his policemen all round our camp at a considerable distance, so as to keep

all intruders from us. The Resident, Mr. Bullock, Captain
Horsburgh and their friends, with great consideration and
courtesy, took their places in or near our living camp; the
army of servants were kept still farther off; and the numerous
natives who had come down by train that morning or the day
before were not allowed to come nearer to us than the village of
Talni itself. We were therefore entirely alone : no one, except
the members of our actual observing party, was in sight, except
our policemen—the sergeant of whom spent the time of the
eclipse at his devotions, and is immortalised in the design by
Mr. Herbert Johnson on the cover of this volume.

I have before mentioned that the walls of our huts were made
of bamboo mats, and we now noticed a very pretty phenomenon.
Through the interstices of the matting, the sunlight fell on the
cement floor in little spots of light. The dark moon had not
encroached far upon the sun before our attention was caught by
a corresponding defalcation in these little wafers of light; and
henceforward we watched the progress of the partial phase as if
by means of hundreds of little pinhole cameras. A precisely
similar effect was noticed by the Resident and his party under
the trees in the dwelling camp, and, as the photograph on
p. 19 shows, it was also noticed at places outside the central
line; this particular photograph showing a piece of pavement
in one of the chief avenues of Bombay.

In the corner of our hut was mounted the same signal clock
—one of those used in the Harton Colliery experiment of Sir
George Airy, but slightly altered for eclipse purposes, and kindly
lent by the present Astronomer Royal—that we had used in our
expedition to Norway. Apparently the heat had affected the
contact springs, for after performing very well from the time it
was mounted till the day before the eclipse it began to give an
amazing amount of trouble, and caused me no small anxiety
during the early stage of the eclipse. Perhaps the greater
coolness of totality had a good influence on it, for during the
two critical minutes it fulfilled its duty without the slightest
hitch.

The falling temperature, as totality drew near, was very
striking. Just as in Norway we had felt the darkness of the
eclipse a sensible relief from the long continuous daylight, so
now in India the coolness and diminution of glare were felt to
be very grateful. The solar topee could be safely discarded, and
we unanimously voted that India would be an ideal land if its
sun could be perpetually half eclipsed.

As the heat declined, the light faded, and the colours died
out of the objects near. The green of the leaves of the trees in
the grove that sheltered our tents turned to a dull lead, and the
blue of the sky was changed into a sombre purple grey.

Then came the end. Mr. Evershed, watching the spectrum,

saw the first indication of the "Flash," and exposed his plate.
A moment later Mr. Smith and I, both of us waiting for the
actual contact, gave the signal that totality had begun. Mr.
Ramrao Subarao, clerk to Mr. Morris, who was taking charge of
the eclipse clock, started it sharply, and as it rang out at each
tenth second called out in a loud clear voice the number of
seconds that yet remained to us. His call of "ninety," "eighty,"
"seventy," the sharp ting of the clock bell, and the subdued
wailing that came from the village beyond the grove, were the
only sounds. The cry of "ten" had gone by but six seconds
when a bright yellow star shone out on the south-west edge of
the moon, wavered for a moment, then spread itself like a
bursting shell, and totality was over.

We took little heed of the remaining partial eclipse. My
wife, indeed, according to our programme, exposed one plate
upon the sun about forty seconds after the return of sunlight;
but, speaking generally, our work was over. Whether it was
done well or ill we had yet to learn. For my wife and myself,
whose chief hope was that we might succeed in photographing
for the first time some of the faint extensions of the corona, we
had marked with much discouragement the great brightness of
the sky during the eclipse. It was an exceptionally bright one,
much brighter than the brightest night at the full of the moon.

The eclipse was over. What secrets our photographic plates
held for us, of course we did not yet know; but we had carried
out our full programmes, and now we became sensible how
severe had been the nervous strain. We might not, however,
yet give way to it. We had to journey down to the railway
station, and for nearly four hours we were standing in the little
triangular tent that had been hastily pitched over the portable
telegraph instrument, beneath a sun that seemed all the hotter
for his recent concealment, receiving and despatching telegrams
about the eclipse, not only to our many friends in India, but
also home to England.

That night Talni village was *en fête*. Throughout the dread
two minutes of totality we were dimly conscious of an undertone
of cries and wailing from the distant village, hidden from our
sight by the trees, and the shout of relief and welcome with
which the villagers greeted the return of sunlight was borne to
us loud and clear ; and now that the crisis was past, the villagers
gave themselves up to unrestrained rejoicings. After dinner
we were invited out to see the illuminations and the fireworks.
Who arranged these I do not know, but they certainly did
great credit to the organiser. Hundreds of native lamps were
distributed about, some on the ground, marking out the paths,
others on frames in long lines of light. This illumination was
arranged on the south-east side of our camp area, that is to say,
at the corner nearest the village. In the foreground, midway

between the spot where our chairs had been put and the lines of lamps, the fireworks were shown. These were simple, but exceedingly effective. They consisted of clay globes filled with a sort of " golden rain " composition, which, lighted at the top, threw out a spout of yellow sparks about 20 feet high. These the exhibitors manipulated with a great deal of ingenuity and taste. Sometimes one or two would be put down flat on the ground, and would spout up their fiery fountains straight into the air. At other times they would be made to cross each other at varying angles, and so with very simple means a good deal of variety was thrown into the exhibition. From time to time also coloured fires were introduced here and there, with which the native spectators seemed exceedingly pleased, but which rather spoiled the effect to us, who preferred the more genuinely native elements of the show. However, we amused ourselves by watching these from our distance through direct-vision spectroscopes, and recognising the familiar lines and bands of strontium and magnesium. Nor were our eyes only to be feasted. Our ears likewise were regaled by the strains of a native band. But here unfortunately our Western prejudices prevented our feeling as much gratitude to the performers as their hearty goodwill deserved. Those who know the party processions in the north of Ireland, and are acquainted with the constitution of a full orchestra in such—namely, five big drums and a cracked fife—will have a correct appreciation of the chief points in the musical programme.

Then came the climax. The great masses of natives moved up towards where we were seated—for not only were the inhabitants of Talni itself out in full force, but the entire *taluc* was represented—then halted some twenty yards away, whilst a deputation, chief amongst whom we recognised the good-looking and courteous Tahsildar of Chandur Taluc, Krishnaji Anaut, came forward, and one of their number addressed us on behalf of the assemblage, bidding us welcome, and asking us to accept the display which we had just witnessed as intended for a token of their wish to do us honour. Then several members of the deputation stepped forward and made the usual presentation of betel and scent, and hung our necks with garlands of flowers; whilst on our side Mr. Smith, as the senior member of our party, returned thanks for us in a very appropriate speech.

This ended the day's work for ourselves, but not for Mr. Evershed, who, directly the flare of the illuminations had faded, began the development of his photographs. The next afternoon, to our great regret, Capt. Molesworth left us, and he was accompanied as far as Pulgaon by Mr. Evershed, who was anxious to compare notes with our neighbours there. Mr. Evershed, who returned to us by the midnight train, brought back a glowing account of the beauty of the photographs which

THE RESIDENT'S CAMELS.

the Government party there had secured. We had already learned by telegram that generally along the line of shadow the success had been unalloyed. Indeed, during the night after the eclipse we had been twice aroused in the deep darkness by the cry, monotonously repeated just outside our tent, of "Tel-e-gram, sah-ib." One of these nocturnal messages informed us, to our great delight, that Mr. Bacon and his party had had, at Buxar, good fortune similar to our own.

Our last days at Talni were busy enough, but need no detailed history. Farewells and packing up were the only incidents. Mr. Bullock, the Chief Commissioner, had left on Saturday afternoon, and waved us a cheery goodbye from the train as he passed the hot little telegraph tent where we were hard at work. The Resident, Sir Trevor Chichele-Plowden, started for Calcutta later in the evening. By Sunday morning half the camp had disappeared. By Monday evening our work at Talni also was ended, and on Tuesday afternoon we took our leave of Mr. Morris, the host to whose untiring thoughtfulness and care we owed so much. Mr. Smith accompanied us for the first stage of our journey, and we parted from him at Nagpur. Mr. Thwaites and Mr. Evershed bade their farewell to the camp the following morning; and when on the Friday we passed in the train for a last time through Talni, our once so busy camp was an empty waste.

With the break-up of our party the report of our eclipse expedition comes really to an end. The rest of our stay in India was no longer in the company of the other members of the expedition, and we were bent on sightseeing rather than on astronomical research. Nevertheless, two or three little incidents had some bearing on astronomy, and it may not be out of place to record them here.

From Talni, my wife and I went to Nagpur, accompanied by Mr. Smith, who bade farewell to us there. At Nagpur we stayed with Dr. Henderson, to whom we owe a heavy debt of gratitude for the perfect rest and quiet that we enjoyed there after the excitement and nervous strain of the eclipse. We had met Miss Henderson in London a few weeks before we sailed for India, and had there suggested to her one or two subjects for observation during the eclipse, which was total at Nagpur. Miss Henderson had accordingly organised a little party of observers, and has very kindly placed their notes in my hands for communication to the Association, and they will be found under the section treating of the "Shadow bands." Miss Henderson herself secured some very pretty photographs of the corona with a kodak.

From Nagpur we went to Cawnpore, and had the pleasure of travelling a considerable portion of the way with Dr. Copeland, the Astronomer Royal for Scotland, on his way home from his very solitary camp at Goghli. At Cawnpore, where we arrived early on Sunday, January 30th, we found that the native mind had been greatly exercised on the previous day by the prediction of some Brahmin that there would be a great earthquake exactly a week after the eclipse. Curiously enough, though there was no earthquake at Cawnpore, yet we learnt later that there actually was one on that morning at Nowshera, and here the patients, native soldiers invalided from the frontier, had such great faith in the prophecy that they had refused to sleep the previous night in hospital.

At Lucknow, at the Imperial Hotel, we met a traveller who was not averse to telling a tale against himself. He had reached the hotel on the eventful morning, January 22nd, and

THE GENTUR MUNTUR, DELHI.

had been shown to his room, with which, for the moment, he felt quite contented. Shortly after he came out in search of the manager, and denounced him in no measured terms for having given him a room that was as dark as a coal cellar. The manager stared at him for a moment, and then—" Why, man, it's the eclipse !"

Agra is memorable to us as the "ascending node" of our orbit with those of several members of the *Egypt* party. Mr. Bacon and ourselves had made many attempts to communicate since his telegram, already referred to, but without success, until our midnight meeting at Agra station, when he entered the railway carriage which we left. This was the first of several such meetings with other members of our expeditions, and five of the Buxar party stayed in the same hotel as we did in Agra.

At Delhi, we, of course, visited the Gentur Muntur or Royal

Observatory, built by Rajah Jey Singh, about the year 1710. This was, of course, long after the invention of telescopes; but the design of the Observatory is essentially that of one where naked-eye work is alone intended, and though not ancient itself, is thoroughly of an ancient type. Just as at Greenwich the transit circle and altazimuth are considered the two fundamental instruments, so here at Delhi the two chief structures were evidently designed for corresponding purposes.

The building that first catches the eye is a huge straight steep staircase leading up to nothing. This is the gnomon of a great sundial, the gnomon being 118 feet in length and its height nearly 57 feet. Right and left from the gnomon are great semicircles on which the shadow falls. South of this structure are a pair of buildings which appear as if intended to be reproductions in miniature of the Colosseum at Rome. In the centre of each is a pillar, towards which are directed from the circular wall thirty stone sectors, and their breadth at the wall equals the space between any two sectors. In the spaces between the sectors are three stories of windows through which the altitude of the star or heavenly body can be observed, its azimuth being given by the number from the meridian of the space in which it is seen. The two buildings are not duplicates, but are supplementary, a sector in one corresponding in azimuth to a space in the other.

Our last station before reaching Bombay was Ahmadabad. Here we stayed with Miss Beatty, at whose house we met Dr. George Taylor, who had seen the total eclipse (at Jeur), and had taken "shadow-band" observations, of which he was good enough to give his report for communication to the Association. Dr. Taylor is also known as being a collector of Indian coins, and as possessing the most complete collection of silver rupees after those at Lahore and Calcutta. This collection he was so good as to show us, and we were specially interested by the Zodiacal rupees of Jahāngīr. These were a freak of that rather erratic monarch, and were produced by him in the years 1027 to 1033 of the Hegira—that is to say, from A.D. 1617 to 1623. The gold mohurs seem to have been chiefly struck at Agra, the silver rupees at Ahmadabad. The selection of the signs of the Zodiac as a design for coinage is extremely unusual, and is rendered more so in the present case by the well-known law forbidding Mahommedans from making representations of natural objects. Jahāngīr was, however, by no means too strict in his adherence to the rules of his religion, for one of his coins bears his own portrait, and, dreadful to relate, depicts him in the act of drinking a glass of wine.

The British Museum, the authorities of which have kindly permitted the reproduction of some of these coins, have a very fine set of the gold mohurs of Agra, all the signs being represented, and several of them by more than one design. I believe that no single set of the Ahmadabad rupees is complete, Virgo, Libra, Sagittarius, and Aquarius not appearing in any of the published catalogues, though rupees showing each of the signs are said to be in existence. Of the others, one of the rarest—the scorpion rupee—is in the British Museum collection, and is one of those figured below.

Our journey to Ahmadabad had been rendered very pleasant by the company of Mr. and Mrs. Oakes. At Bombay we met Mr. Henry Cousens, who had taken so much trouble on behalf of the party when it was hoped that we should be able to arrange a camp at Masur. Mr. Cousens had observed the eclipse himself from Jeur, and had taken six beautiful negatives of the corona, which he presented to my wife, besides a number of photographs of the camps of our fellow-members, Prof. Naegamvala and Prof. Burckhalter, and also of the Lick Observatory party under Prof. Campbell.

E. WALTER MAUNDER, Secretary.

ZODIACAL RUPEES OF THE EMPEROR JAHĀNGĪR, MINTED AT AHMADABAD.
(The eighth coin shows the inscription on the reverse.)

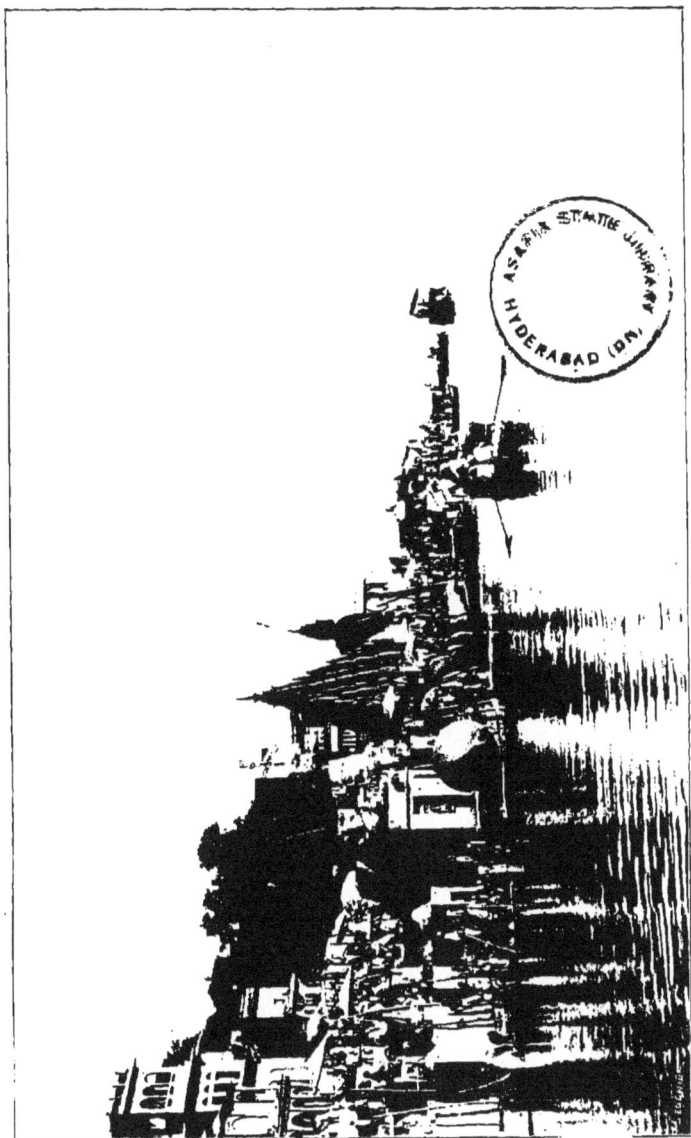

BENARES.

CHAPTER III.

THE EXPEDITION AT BUXAR.*

IT must be confessed that when, on December 8th, I parted with my co-secretary on board the *Ballaarat*, and wished him and his party God-speed, things appeared anything but promising for the two British Astronomical Association Eclipse Expeditions. With one fell stroke our long discussed and carefully laid plans had been swept to the winds, and we found ourselves at the moment of departure scarcely more advanced in our arrangements than we had been a year previously. In the fortnight that elapsed between the sailing of Mr. Thwaites's contingent and our own no fresh developments occurred. Indeed, this was hardly possible, and when on December 23rd we embarked at Tilbury on board the R.M.S. *Egypt* we had but the vaguest ideas as to how our future course might shape itself.

We left England a party of ten members—Miss Dixon, Mr. W. B. Gibbs, Mr. E. W. Johnson, Mr. H. Keatley Moore, Mr. and Mrs. Nicolson, and Mr. F. Lys Smith—besides my own family party. We were joined at Marseilles by Mrs. and Miss Bevan and the Rev. S. Hebert, and being further reinforced at Brindisi by Mr. and Mrs. Oakes, our number attained its full complement of fifteen.

Our start from Tilbury can hardly pass without mention. A week or more of heavy fogs culminated with one of exceptional denseness on the day of our departure. The railway traffic was seriously disorganised, and the special from Liverpool Street did not leave till long after its appointed hour; though this was of small consequence, for, arrived at the river, we found an impenetrable wall of fog, through which the tender pushed its way to our ship, lying in mid-stream and absolutely invisible to us till its great black hull loomed dark above our heads.

We were advertised to sail at noon, but as the day wore on the fog continued to increase, and advices reaching us that it was thicker yet down stream, we were perforce obliged to remain where we were, enveloped by a pall of mist through which the bells

* By the Rev. John M. Bacon, *Secretary.*

of the surrounding shipping tolled dolefully through the night. Later on there was trouble. A vessel near us broke her moorings and went adrift among the other shipping, till she presently filled and sank, her crew escaping with difficulty; and at day-break we saw her, half a cable's length astern, the tops of her masts alone standing up above the turbid stream. Other disabled craft lay half submerged or beached, while one large steamer had fouled ourselves during the night, and was lashed to our side for mutual protection. It was a welcome relief when at eleven o'clock of the forenoon of Christmas Eve the fog cleared away as if by magic, and in bright sunshine and under a blue sky the *Egypt* stood down stream for the Channel.

This was perhaps scarcely a propitious commencement of our voyage, but whatever evil fortunes it might seem to portend were unfulfilled, and henceforward the good luck that attended our course began to manifest itself.

The weather throughout our voyage was all that could be desired. We had a cold spell to begin with, when those who sat on deck needed the protection of ample wraps; but this state of things lasted but a short while. The Bay was merciful to us, and, once beyond the frowning fortress of Gibraltar, we found ourselves in warmer climes. In our boat we were excep-tionally fortunate, the *Egypt* being at that time the newest and finest of the P. & O. fleet, and in point of speed and comfort leaving nothing to be desired.

With assured weather and completed numbers, business began to employ the time of our party for a recognised period of each day. As a preliminary step a general meeting was held in order to review the different instruments available for the observation of the eclipse, and to assign the various duties to each member.

The peculiar circumstances of our expedition, its uncertainty, and the shortness of the time intervening between our arrival in India and the day of the eclipse, had of necessity compelled us to abandon all instruments and apparatus of a cumbrous nature, or such as would require many hours to erect and adjust. This was in many ways a sad loss, though unavoidable; but our equipment was at least as complete as was compatible with portability. The heaviest and most important instrument was undoubtedly the animatograph telescope, specially designed for the expedition by Mr. Nevil Maskelyne. As a piece of delicate and complex mechanism this apparatus demanded much care and manipulation, and a certain period of each day was set apart for requisite practice. At intervals, too, during the voyage, a practical use was made of its capabilities. A number of spare films had been supplied by Mr. Maskelyne, in addition to one of special length for the eclipse; and with the consent and ready co-operation of our genial captain several photographs of animated scenes on board ship were secured.

The $4\frac{1}{10}$th-inch telescopic camera designed and manipulated by my son was, next to the animatograph, the largest piece of apparatus belonging to our equipment, and remained buried deep in the hold during the voyage. Mr. Hebert, however, was possessed of a fine $4\frac{1}{4}$ Cooke refractor, which, being more portable, he was able to erect on several occasions. The cameras, large and small, owned by Mr. Smith, Mr. Oakes, Mr. Hebert, and my daughter did excellent service on very many occasions before the eclipse, for which event they were primarily intended. From an early date Miss Dixon commenced careful practice with her special instrument—an opera-glass fitted with a slitless spectroscope, lent to her by Miss E. I. Page,—a piece of brown paper, covering a porthole and having a narrow opening, yielding an excellent object for preliminary trials.

To the very able care of Mr. Moore was entrusted the responsible task of organising and training a sketching party to draw the corona as seen with the naked eye, and under his direction most regular and praiseworthy work was carried out. Almost every evening he and his indefatigable band assembled in a lower saloon with chalk and paper; when a drawing of the corona was fastened at the proper elevation and exposed to view for ninety-six seconds only, during which time each operator made a drawing of the quadrant allotted to him, the whole sketch being then put together and compared with the original. These practices were a source of much interest, and by their means considerable proficiency in rapidly grasping the outlines and correctly committing them to paper was attained.

As more southerly skies were reached, the heavens, now unfamiliar to most of our party, afforded us ever fresh delight. Night after night new stars and constellations began to reveal themselves; and while eagerly identifying objects we had hitherto known only by name, we were equally interested by the unwonted aspect assumed by our old friends of the northern skies. Orion stretching prone across the heavens, the Great Bear with his tail now dipping in the water, were sights as novel to us as Canopus or the False Cross. The true Southern Cross, of course, was not visible till after midnight; but few of us will forget our delight at seeing for the first time that famous constellation of the southern heavens. Nor was it long before the Zodiacal Light began to claim general attention. Each evening during the latter part of our voyage the cone of ethereal light appeared to stretch itself farther and more clearly across the sky; and to watch for its first appearance, as also for the shadow of the earth rising from the placid sea, was an unfailing pleasure. The lunar eclipse of January 7th occurred during our second night in the Red Sea. Anything approaching careful observation was of course out of the question; but to the credit of our party be it said that scarce a member failed to put in an appearance on the boat-deck in the small hours of the morning and watch the black shadow eating its way into a moon that shone with almost blinding brilliance.

We flattered ourselves, and not without reason, that the leaven of fifteen members of an eclipse party imparted a distinctly astronomical tone to the whole of the passengers of the *Egypt*. A spirit for sky-gazing animated the entire ship, and certain astronomical literature with which both Mr. Gibbs and myself were provided was eagerly devoured by passengers of both saloons. Indeed, I believe that Mr. Gibbs had considerable difficulty in reclaiming a copy of Mrs. Todd's *Total Eclipses* at the end of the voyage, into so many different hands had it passed. This general interest in our mission had also the happy effect of procuring us some valuable aid from several Anglo-Indians among the passengers. Specially were we indebted to Mr. H. B. Beames, of Calcutta, whose knowledge of India, and particularly of the district we were bound for, enabled him to give us most useful information and advice; while from Mr. W. Sargeaunt, of the S.M. Railway, we gleaned invaluable hints relative to the natives and customs of a land we were all visiting for the first time. Both these gentlemen were good enough to attend some of the many meetings held to discuss our future arrangements.

We woke up on Friday morning, January 14th, to find our boat slowed down and creeping in with the tide into Bombay, and then it was that we woke also to the reality of the work before us. It wanted but eight days to the eclipse; yet we did not know if any provision was being made for us, nor were we even certain of our destination. We simply knew that our party was to muster on the shadow track at some point to be determined by circumstances, and by the facilities offered by our chosen line of rail, but which, at any rate, could not be less than a thousand miles inland. But further than that our general instructions availed us nothing. Indeed, the suggestion originally made to us that we should take up our quarters at Benares or Allahabad, and go down to some observing station on the morning of the eclipse, appeared to me all along a most doubtful proceeding, and, as will be shown, one which would have proved a sheer impossibility. So firmly convinced was I myself of the necessity of trying to run a camp, at all hazards, that before leaving home I had written round to our whole party asking who would be prepared to join in the venture. About one-half only gave in their names; but so greatly had enthusiasm been stimulated during the voyage that now, at the end of it, every member of our expedition was keen for any sort of real campaigning experience, if it were only possible in any way to place ourselves under canvas.

MR. AND MISS BACON.

MR. FRED BACON.

Towards this end I had also, before leaving home, written forward to a number of hotel contractors, English residents, and also Government officials, bespeaking their help and advice, and begging that their replies should await the arrival of the *Egypt* at Bombay.   Thus it came about that no sooner had we dropped anchor than we were boarded by a boat from shore; and five minutes later my steward brought me nearly a score of letters, half of them official, and nearly all requiring immediate answer —and for a good reason.

For now it was that we learnt that the Viceroy, likewise the governors and potentates of one quarter of India, had elected to

THE COOK AND HIS KITCHEN, BUXAR.

observe the eclipse from the very centre which we ourselves had come six thousand miles to occupy.   This necessarily meant an enormous gathering, with the levying of every tent in the district and of all things essential to camp life.   This was really not the worst trouble before us, but happily at that moment we were not aware of the fact.   Private individuals had all written to tell me that nothing could be done, and moreover that there was no available accommodation near any observing station.

However, as this adverse tide of circumstances swept over us, I clung desperately to two huge official envelopes, one from District Magistrate Mr. E. P. Chapman, conveying in the most generous terms an offer from Commissioner J. A. Bourdillon, of the Patna Division, to aid us to the extent of his ability, the other from the local magistrate, Mr. J. D. Cargill, marked by the same unbounded courtesy, but frankly explaining that the resources of the neighbourhood were after all limited, and might well prove insufficient for our requirements.

But ere these letters had been fairly read, our vessel lay off the Ballard Pier, the tender was alongside, the gangway run out, and the moment was already come for a hasty yet most hearty leave-taking of our captain and officers, who had been such true friends for three happy weeks. Five minutes later we found ourselves contemplating long groves and avenues of piled-up baggage, and verily in a strange land. We were thrown entirely on our own resources; not a soul of our party had been in the country before, nor knew the native or his tongue, nor had we one friend in all Bombay to turn to. Worse still : as all credentials were in my keeping, it was essential that I should go forward and with all speed make due arrangements with the railway company ; yet the heaviest of the instruments and baggage were my own, and half of these lay buried somewhere under chaos, and seemed likely, for all that we could do, to remain so for good and all.

Of course in the end things righted themselves somehow. A Babu came to my rescue, unearthed my bulky belongings, and while I cleared all at the Customs, engaged a small army of coolies, with sundry gharries and bullock-waggons, and then— asked for payment. How many rupees I had to overpay him I shall never know ; yet the money was well invested, for it bought me experience which was simply invaluable. However, at the railway station, on the arrival of the bullock-carts, there seemed to be every one of those coolies again who had been already overpaid, and a vast number more besides, all salaaming and clamouring for more backsheesh. It was a Parsi this time who stood my friend, promptly dismissed the rabble, and booked our baggage at an outlay that by comparison must be deemed moderate ; and to the end of my days I shall remember that courtly gentleman with his shiny hat with the utmost gratitude.

My own family party, accompanied by Miss Dixon and Mr. Johnson, started forward that night for Benares, as pioneers for the main body of the expedition, who elected to remain in Bombay and await the further development of plans. Messrs. Gibbs, Hebert, and Smith also journeyed up by the same train. And so 10 p.m. found us, with the satisfaction of feeling that not an hour had been lost, steaming out from under the heated roofs of the Victoria Terminus, the fresh air from the open country blowing through us, and a journey of forty hours on

The increasing speed and swing of the train, though not always easy, was inexpressibly soothing to nerves somewhat tried by the unwonted heat and bustle of an anxious day and impatient of all hindrance ; and it was delightful to watch the novel, weird-looking country, dimly outlined in the tropic starlight, sweeping past us as the night mail settled to its work. But ere midnight there was a check ; some block was on the line, and we pulled up somewhere in the jungle without assignable cause or means of obtaining explanation save in a heathen tongue. The annoyance of an irksome delay was however presently relieved by the arrival of another train, which drew up alongside of us. There were English tongues here,—no, Irish : five hundred of them ; half a regiment of Inniskillings, mainly raw troops, were on their way to the front ; and a dozen red-coated lads craned their heads out of the train, delighted at the opportunity of a chat and the exchange of a little good-natured banter. They told us tales of their homes away in Cork, of incidents of their voyage, and of their prospects of rough and active service ; and this last we learned to appreciate a little later when we came up with another train—meeting us this time, and bringing down from the frontier a sorry band of old campaigners, sick and wounded and worn with the severity of their late service.

The only other incidents that broke the monotony of the long ride were the regular stoppages for medical inspection. A solemn farce as far as we were concerned, at first somewhat amusing by its absurdity, but growing terribly tedious by its repetition. During one night halt there was a hunt up and down the train for the addressee of an official telegram, which was presently handed to myself as being the individual concerned ; and then I learned how my colleague Mr. Maunder, the indefatigable senior secretary of the eclipse expeditions, had even in the thick of all his own anxious work found time somehow to make interest with the authorities on our behalf, and to aid materially in setting matters in train for our benefit.

Owing to the many delays we missed our train at Moghul Serai, and so found ourselves at that comfortless junction at 5 a.m. of the third day, much jaded with so long a spell of travel, and sorely in need of warmth and breakfast. It was then, as on so many like occasions, that the ladies of our party showed their true helpfulness and ready resource. They now took the lead. Down in the depths of our travelling kit were a spirit kettle, a teapot, bovril, biscuits and such-like fare. These we unearthed, and being in strong force, next advanced upon an empty ladies' waiting-room and promptly carried the position. But where was the pump ? The station-master soon settled this difficulty. Kellner's refreshment rooms would not be open for hours, but he got us access to the filter, and—we were soon feeling better. Perhaps it was the most blissful hour in our lives when four hours later we reached Clarke's paradise of an hotel at Benares and the luxury of an Indian bathroom.

A little later, tiffin over, the spirit of curiosity began to assert itself, so by general consent gharries were ordered, and our entire party started off for our first visit to the native quarter. The ancient observatory was our first goal, and naturally possessed special interest in our eyes; but we were disinclined for much exertion in the heat, and quickly yielded to the temptation of a river barge. And so we were soon drifting past those sights of the ancient Brahmin capital surely unequalled in all the earth: the endless piles of crowded temples, the huge and hideous gods, the worshippers, the devotees, and the nameless horrors of the Burning Ghât; the memories of all which can surely never more fade out.

But the time for sight-seeing was not yet. That night Mr. Johnson and myself, dodging the mosquitoes somehow, got two or three hours' beauty sleep, and then rose at 3.30 to catch the early train to Buxar, our chosen eclipse ground, fifty miles away. A hasty "chota hazri" and a sleepy "gharry wallah" were in waiting, and we were at Benares station an hour and a half yet before sunrise. Of course it was pitch dark, and the babu deputy station master was not there, but his office was open and we entered to have a worry round. The place was bare, however, except for a number of dark-coloured sacks neatly tied up and arranged all over the floor in orderly double row. Through lack of light I trod on one of these, which spilt me over a second, and so spread-eagle fashion on to half a dozen others. The sacks shook themselves, mildly remonstrated, and settled down again. It was pilgrimage time, and these were Hindus.

At Buxar station we were met by Mr. Chapman and Mr. Cargill, who gave us a most cordial welcome, and as a first proof of that unbounded and, in my experience, unexampled hospitality which followed, treated us to a breakfast at Kellner's, the like of which is unknown at any English railway refreshment rooms. Then we were conducted amidst much serious talk to a solitary mango grove, a short half-mile distant, and there we learnt how severe a tax was indeed being put upon the whole country-side to supply canvas, provisions and necessaries of all kinds for the distinguished guests who were coming with their legions of attendants. Moreover, as I have mentioned, the great native pilgrimage of the year was in progress, and a ceaseless tide of dusky humanity from all over India was setting along every highway, blocking the railways and crowding every road with the million and more that were gathering to bathe in holy Ganges on eclipse day. And that now was but five short days hence. Before half the tale was told it seemed as if our venture must after all end in disaster. Surely under such

THE CAMP AT BUXAR.

Mr. Hebert.   Mr. Lys Smith.   Mr. Gibbs.

A GROUP OF OBSERVERS.

circumstances nothing but the fabled magic of an Arabian tale could conjure habitation on to that spot in the time.

But as yet I knew not India nor the power of a Commissioner and his Deputies. I could only read in the thoughtful faces of our hosts that the task they were so ungrudgingly undertaking on our behalf was little less than Herculean, and involving, as I had reason to fear, much personal sacrifice. I knew only too well that this was so, three days after, when I stood in the same mango grove surrounded by an imposing array of Indian tents, all the equipment of camp life necessary and sufficient for our whole party, an ideal observing ground in the centre, and two clear days yet to the eclipse!

OUR FIRST MEAL AT BUXAR.

The original pioneers mustered on the ground in the course of the day, and the accompanying illustration shows our first meal *al fresco*. Our provisions at that period were likely to have been scanty, but a royal birthday cake had sprung from somewhere—soon proved to be the thoughtful gift of Mr. Chapman himself, who, moreover had engaged our necessary staff of servants and had even sent us his own native cook, that moment busy in our kitchen. At the far end of the enclosure stood a noble Swiss cottage dining tent, which, as also much of the camp furniture, I sadly fear was simply the private property of Mr. Cargill.

But half the charm of all lay in the fact that we ourselves should have a share in arrangements as yet necessarily incomplete, and that in a small way we were to make our bivouac on the Indian plains. It only added to the fun that a bed or two had that night to be laid on clean straw, and to find next morning that there was as yet but one basin to go the round of the party. We soon got things into shape, and ere the main body joined us, after tiffin the next day, the camp was in full swing; a dinner of many, if unknown, courses was served in the mess tent, and that night the entire expedition slept soundly on their own ground through the strange silence broken only by the bark of the jackal in the bush jungle and the practised cough of the party of "chowkidars" who mounted guard over us.

The morning of the eventful day broke fine and clear. A few days previously some anxiety had been aroused by the alarming tidings that the customary three days of Christmas rain had been unaccountably delayed, and might yet be almost hourly expected; and when, for two or three evenings, ominous clouds had begun piling themselves against the sky at sundown, forebodings of a repetition of the Norway disappointment perforce entered our minds. But our fears were groundless, as one glance at the pure blue sky in the early hours of Saturday, January 22nd, sufficed to show. We were to see the eclipse on *this* occasion, and the camp was astir betimes.

A dress rehearsal of the eclipse programme had been held the previous day, at the hour when the sun reached that position in the heavens which it was calculated to occupy at the period of totality; observers therefore could afford to prepare for their several tasks in a leisurely and thorough manner, and we were free from all flurry or fatigue. For all this, considerable excitement ran round the camp when Mr. Hebert, whose eye had for some time been glued to the small end of his telescope, presently looked round and announced "First Contact."

About this period, also, visitors to our camp began to arrive. Col. Strahan, an old friend and very talented member of the Norway eclipse expedition, had already come to offer any needed help, and now our observing force gained three other valuable allies. Col. Sinclair, R.E., having read over the syllabus of work that had been mapped out for our station, elected to make observations and sketches by aid of a fine field-glass from a neighbouring lofty semaphore stage; Mr. Willoughby Meares, F.R.A.S., also joined forces, while Mr. Cargill undertook the task of watching for shadow bands on his own *pukka* tennis-court, admirably adapted for this purpose. A note arrived from Lord Rosse inquiring for accommodation in our camp; and next Mr. Babbington-Smith, the Viceroy's private Secretary, called in person inviting me to lunch with his Excellency after the eclipse, and to take an animatograph picture of the brilliant gathering that would then be assembled. Representatives from the Indian Press and from the native observatory of Madras

also presented themselves, and ere these small distractions were
over the Partial Phase was far advanced. But the animatograph
was already in place and duly oriented. The several cameras
presided over by Messrs. Oakes, Smith, Hebert, and my
daughter, might all be relied on to secure their due records.
My son, though unaided, had, as the result proved, arranged his
$4\frac{1}{10}$ refractor to take two perfectly successful photographs of
the commencement and middle of totality. Mr. Gibbs, Mrs.
Nicolson, and Mr. Johnson were to take general observations.
the latter undertaking specially to record the appearance of
shadow bands, and to expose a plate prepared by Mr. Gare, and

PILGRIMS WAITING NEAR BUXAR STATION.

designed—equally with a somewhat similar device of my own—
to compare the light of the corona with that of the full moon.
Miss Dixon, assisted by Miss Bevan, gave her undivided attention
to the slitless spectroscope; and it is needless to add that Mr.
Moore's sketching party, including Mr. Nicolson, Mrs. Bevan,
and Mrs. Oakes, were to be seen seated with bandaged eyes
ready for the moment of totality. This I was able to announce
myself, standing at my animatograph with a metronome
ticking seconds beside me. I was also able to count off aloud
each tenth second throughout totality, keeping time the while
with the smooth and perfect motion of Mr. Maskelyne's beautiful
instrument.

At the expiration of 92 seconds the total phase abruptly
ended, the light came flooding back, and a few moments later a

wild shout reached us from the tens of thousands of Hindus gathered on the banks of the Ganges a mile away, telling that they too had realised that the great eclipse was over.

A formal meeting was held that night, which Col. Sinclair also attended, when a general report was drawn up; and the next day saw the break-up of our party, bent on pursuing various routes of exploration through the country. My own work, however, was as yet not over. My report had to be written, the accounts made up, and the staff of servants paid off, which necessitated my staying one more night in camp with my son and daughter. And here I must own that I was confronted with a serious difficulty. An unaccountably large number of natives mustered, all bewilderingly alike, and each and all maintaining that they had waited on us in the camp and clamouring for payment. Moreover, the few who had professed to talk a little English now found that they could not understand the simplest question.

In this dilemma I was constrained once more to fall back on that never-failing aid that had brought us through every difficulty. I promptly dismissed our chief servant—much to his chagrin—with a note to Mr. Cargill as local magistrate, explaining matters; and by morning light his own private secretary came over to the camp, and in a very brief space of time the whole trouble was over. The impostors slunk off. The actual staff received only their just payment, and our most kind and courteous friend declined to leave us until he had personally seen the full tale of our belongings delivered at the station.

From here, following up the route originally proposed for the expedition and timed to gather up all home letters at pre-arranged points, we visited in due course the historic battle grounds of the Mutiny, the Himalayas, and Calcutta, fore-gathering everywhere continually with detachments of the general B.A.A. force.

A few places of principal interest will be found among the illustrations.

JOHN M. BACON, *Secretary.*

CHAPTER IV.

THE CAMP AT JEUR.*

AMONG the many convenient sites for observation of the total solar eclipse of January 22nd last which lay along the central line of the belt of totality, Jeur, a station on the Great Indian Peninsular Railway, presented many points of advantage. The more favourable sites as regards the length of the period of totality, in the Satara district, being within the area of plague infection, had very reluctantly to be abandoned. Indapur, though south-west of Jeur, was but twelve to thirteen miles distant, and the very small difference that would have been gained in the period of totality would have been so small as to make the great disadvantage of the distance of the former place from the nearest railway station—thirteen miles, with an indifferent road and a river to cross—outweigh it entirely. But even Jeur, owing to the bad name the plague had got, was given a wide berth by the great majority of observers. A few only of the more daring spirits ran the risk, which was practically *nil*, and were rewarded by an appreciably longer series of observations than those in the Central Provinces and farther north.

The first to arrive on the ground, six or seven weeks ahead, was Professor Campbell and party, from the Lick Observatory, who, after searching the country in vain for some distance around for a hillock to rest his long telescope against, settled down upon a spot of high ground about four miles south of Jeur station. Next came Professor Naegamvala, from the College of Science, Poona ; and a Japanese party from Tokio University under Professor Terao, the former discarding his previously reserved areas north of Jeur station for a plot close beside the Lick party. Lastly came Professor Burckhalter, from the Chabot University, Oakland, California, with the Pearson telescope. With some of the professors from the College of Science, Poona, to help Mr. Naegamvala, Dr. Arthur Thomson, and an amateur or two, the eclipse camp was complete, and the time from the arrival of the various parties up to the morning of the eclipse was fully occupied in unpacking and

* By Henry Cousens, Superintendent of the Archæological Survey of India. Communicated by E. Walter Maunder, F.R.A.S.

4

setting up the instruments, adjusting, sighting and focussing them, making trial plates, and drilling the assistants. The last was not only done during the day, but also at night, to accustom them to work in the dark. Our American friends rather astonished the natives, for there was nothing they could not do, and did not do, with their own hands—from driving in a screw or sawing a plank to the most delicate adjustments of their instruments. Visitors might have passed them by at their work in looking for more presentable professors. Having brought their own tools and implements, they were certainly independent of the village carpenter.

The central line of the eclipse was shown upon some maps as passing through Targaon, on the S.M. Railway, and Jeur, as nearly as possible; but the American parties were not satisfied with this alignment, and pitched camp farther eastward, they having concluded that it would cut the railway line between Jeur and Kem, about four to five miles from the former place. Professor Campbell's camp was within two miles north-west of this line. Adjoining his camp, but farther from the line, was Mr. Naegamvala's, while the Japanese were a hundred yards farther on again. Professor Burckhalter was at the village of Vangi, about five and a half miles from Jeur and one mile and a half beyond the other camps, but on a parallel line with the rest so far as distance from the central line was concerned.

The Lick camp was conspicuous by the long 40-foot telescope, which, with its end planted in a pit some 8 ft. deep, shot up into the air above the tree-tops. A marked feature, too, was the Stars and Stripes which, supported by the Union Jack, very appropriately floated above the *chupper* roof of their banqueting hall. It was a sign of good feeling and comradeship which characterised the whole intercourse between the camps. In the shadow of the gigantic tube, and looking quite dwarfed beside it, were the Professor's other instruments for spectrum photographs and smaller coronal images. The great telescope gave a disc of nearly 5 in. in diameter, the largest obtained in the camp. But it was interesting to note that the astronomers were not too proud to use the humbler and smaller everyday camera for snapshots—all were pressed into service.

Professor Naegamvala's largest and most important instrument was a large objective prism mounted upon a very heavy substantial stand, upon which it moved by clockwork. Another instrument for spectrum photography worked with others off the cœlostat, whilst an integrating spectroscope, working direct, was manipulated by hand. A telescopic camera giving an image off the cœlostat nearly an inch in diameter, an eye spectroscope, two smaller telescopes, a small camera, and thermometers of sorts, with wire-encircled discs mounted at the top of poles for drawing the corona, swelled the number of his

Miss MacBean.   Mrs. Campbell.   Prof. Campbell.

THE BASE OF PROF. CAMPBELL'S GREAT TELESCOPE.

instruments of observation, which made indeed a brave show. Some of these only just arrived in time from England.

The Japanese contented themselves with less in the way of apparatus. A long bellows camera working off the cœlostat, which was of a different pattern from Professor Naegamvala's, and apparently more complicated, had its other end fixed against the window of a little portable dark-room which the party had brought with them from Japan; and the exposures were made with a sliding shutter. With this they obtained a disc about 4 in. in diameter. A short focus telescopic camera in addition, moved by clockwork, gave them another set of photographs of the corona on a smaller scale. They did no spectrum work. All their arrangements and observations were carried out by themselves. They had little outside assistance save for common coolie work. They brought their own carpenters with them.

Professor Burckhalter, in his little nook among the trees by the side of an old garden well, just outside the town of Vangi, on the north, was simpler still in his apparatus. He intended doing little, but doing that well; and, judging from his results, he was in this eminently successful. His only instrument was the Pearson telescope, which he duplicated by having a second lens of the same make and power, mounted in a home-made tube, clamped on to the top of it, the two working together. These gave him images about 2 in. in diameter. By a very ingenious device of his own he was able to control the exposures of one set of his plates, so as to give the outer streamers a long exposure and the chromosphere a shorter one upon the same plate. This was done by a revolving disc in front of the plate, with a slit cut out of it wide at the circumference and less at the centre. The disc was revolved by clockwork at the back of the plate, the attachment passing through a hole made in the plate itself. This hole falling in the centre of the moon's disc did not matter much, but its effect upon a print would be uncommonly like some newly disclosed cavern or empty crater upon the moon.

Professor Naegamvala, who had by far the greatest number of assistants upon the ground, was helped by the Principal, some of the professors of the College, and other gentlemen, and a strong contingent of native students whom he succeeded in knocking into shape by constant drills for a week in advance. Helpers for Professors Campbell and Burckhalter arrived on the ground only a day or two before the event, and were recruited from the British Naval and Military Departments at Bombay. The Japanese depended entirely upon themselves, and they were an ample party for the few instruments they were using.

The day of the eclipse wore on, anxious hands adjusted and readjusted the instruments, the drilling of the assistants was more frequent and thorough; and the astronomers, by this time worn out with work and anxiety, awaited the momentous event with trepidation and fear lest something should happen to thwart all their expectations, some little cloud arise, some ill-disposed breeze blow. Day after day for weeks together not a cloud had been seen in the sky; it seemed almost too much to expect this state of things to last much longer. But if it would only keep so until the 22nd! For days previous to the eclipse the wind rose in strong gusts about midday, and there was at times a strong tremor in the air due to the heat. Great wind-guards of sacking and bamboo rose to windward of the larger instruments, and each day saw some further arrangement or addition to these as the continual recurrence of this wind gave rise to greater anxiety. The basement of the great wooden framework carrying Professor Campbell's big lens was firmly embedded in a great heap of stones piled around its four sides.

The morning of the 22nd broke as usual with a clear sky save for a low-lying smoky-looking haze along the western horizon, but as the sun rose this was dissipated. The astronomers were early astir scanning the horizon, and as the morning wore on hope was fast turning to certainty. Towards noon, however, that tiresome wind began as usual; and then, began too, a tussle between it and the eclipse, but as the latter advanced the former after several ineffectual attempts to work mischief gradually died away in fitful gusts, and during the two minutes of totality there was a perfect lull. Even the tremor in the atmosphere, probably due to the cooling down of the earth, was less than it had been.

The beginning of totality was not so marked as was expected. No wall of shadow was seen to sweep over the landscape. The disappearance of the last ray of light left it rather darker than before, but not dark enough to necessitate the use of lanterns in making observations; in fact, the amount of light during totality was remarkable, the astronomers all being struck by it. The two minutes during which it lasted hardly seemed so long to the observers, who could barely press all they had planned into the time; and the first reappearing shaft of sunlight found some in the midst of a last exposure. The tension of weeks snapped, anxiety was at an end, the astronomers had secured their results, and the sun gradually made his reappearance, but his subsequent proceedings interested them no more.

Though arrangements had been made to watch for the shadow bands, they appear to have been hardly noticed; but near Indapur, where two ladies had spread a sheet upon the ground, they were well seen both before and after totality. Their attention was first attracted to them by what at the first instant seemed like the shadow of a fluttering bird overhead. The

PROF. NAEGAMVALA'S CAMP, WITH PROF. CAMPBELL'S GREAT TELESCOPE IN THE DISTANCE.

bands appear to have been about three to four inches wide, and followed each other in wavy lines across the sheet. One of the assistants at the camp was told off to watch the effect of the eclipse upon the animal world, but unfortunately the only animals in sight were a couple of miserable *tonga* ponies, who displayed a sad want of appreciation of the event, for they had not a soul above their hay. The corona, though I did not see it myself, being too busy with my photographs, was said by all to have been particularly colourless; it was described as a silvery light.

Whilst serious work was thus going on at the eclipse camp, Jeur station and its surroundings were rapidly putting on a holiday appearance. A month or so previously some one suggested that the railway company should run a special from Bombay to Poona, which, after some hesitation, they advertised to start if a sufficient number applied for seats beforehand. Not one, but eight long specials steamed into Jeur on the eclipse morning from Bombay, Poona, Ahmadnagar, Hyderabad, Baroda, and Madras, and emptied out hundreds of European and native visitors, some keen on observing the eclipse and the astronomers, others out more for a picnic and the fun of the thing. One part of the programme of the former was seriously interfered with by the police, who had formed up across the road leading to the eclipse camp, and stopped all excursions thither until the crisis was over. The observations were thus made in the greatest comfort, and the camp was absolutely free from intrusion. Scores of white tents crowded around the station and peeped out from the surrounding trees. The goods shed was converted into a great refreshment-room, and covers were set in first-class style for three hundred, but the majority of the visitors had brought their own tiffin baskets. The shed was gaily decorated with flags, bunting, evergreens and Chinese lanterns. A special Government telegraph office was the only sign of business. Jeur had never before seen so much of the Western world, and, if the astronomers are correct, it will be a very long time ere it will see it again.

HENRY COUSENS.

CHAPTER V.

SPECTROSCOPIC OBSERVATIONS.

THE object of observing a total eclipse of the sun is of course to increase our knowledge of those surroundings and appendages of the sun which are then, and then only, revealed to us. So intense is the glare surrounding the sun on ordinary occasions, from the lighting up of our atmosphere in his immediate neighbourhood, that the whole of the beautiful and complicated objects which form, as it were, the pavilion in which he perpetually abides, are entirely hidden from us in ordinary daylight. Though very bright themselves, they are drowned in the far greater brightness of the atmospheric glare. And this glare renders it impossible for us to bring the sun's surroundings into view by arranging a kind of artificial eclipse ; or else it would be easy enough to form an image of the sun and its appendages in the focus of a telescope, and then, hiding the sun itself behind an opaque disk, to examine its neighbourhood at leisure.

Up to the eclipse of 1842, July 8, very little notice had been taken of the revelations which solar eclipses afforded of strange and beautiful objects apparently surrounding the dark body of the moon, and which might belong either to her, or to the sun which she at such times concealed for a brief interval. But in that year the shadow track passed over the entire length of Europe, and all the first astronomers of the day were induced to take part in its observation, and the attention of every observer of that eclipse was caught by the beautiful rose or ruby coloured lights of irregular forms and distribution, which were seen round the dark moon. From that time forth, "the red-flames," "prominences" or "protuberances" as they have been variously called, have been as thoroughly enrolled amongst the objects of astronomical interest as comets or nebulæ.

Nine years later the shadow track of another eclipse, that of 1851, July 28, passed across Europe, but from north to south, not west to east as in 1842, and the prominences received yet further attention. It was then made clear, as the acuter observers in 1842 had inferred, that they belong to the sun, not to the moon ; and it was also seen that they rise up from a thin shell of similarly coloured material, since known as the "chromosphere," which covers the sun pretty uniformly everywhere to a depth of about 3000 miles, whilst the prominences throw their weird fantastic branchings sometimes to heights of 100,000, or it might even be of 150,000 or 200,000 miles.

But beside the prominences, these eclipses had caused the recognition of a more astonishing object still : an irregular glow of silvery light which spreads out from the sun like a vast star, in all directions, and which has been known to us since as the "corona." From that time forward the form and details of the corona have been drawn and photographed upon every possible occasion; though from 1851 up to the present time the entire duration of all the available opportunities has not amounted in the aggregate to as much as two hours.

No great progress was made, or was indeed possible, until the spectroscope was turned upon the prominences in the eclipse of 1868, August 18; like that of last year, visible in India. It was at once seen that they gave a spectrum of bright lines—the unmistakable sign that they were composed of glowing gas—and these lines were so brilliant that M. Janssen found himself able to observe them after the eclipse had passed away. The chief lines were speedily seen to be those of hydrogen gas; others remained unassociated with any known terrestrial element until 1895, when Prof. Ramsay discovered "helium," the source of most of them, in the Norwegian mineral "cleveite."

From the time that Janssen and Lockyer succeeded in observing the prominence lines without an eclipse, and so brought them within the range of daily observation, the interest attaching to prominences as parts of the phenomena of an eclipse passed away, and the chief questions which eclipses were looked to to answer were connected with the corona. The answer of the spectroscope as to its character came more slowly than in the case of the prominences, for the corona was a more complex structure and its spectrum showed a corresponding want of simplicity. The coronal light appears in effect to proceed from three sources. Part of it is reflected sunlight, and this shows itself in a feeble reflection of the solar spectrum—a continuous spectrum, that is to say, crossed by the dark lines, known by the name of their discoverer Fraünhofer. Part is due to the glow of intensely heated particles of dust or solid matter, and shows a continuous spectrum free from the Fraünhofer lines. These two sources are naturally not easy to discriminate the one from the other. The third source is glowing gas, giving a spectrum of bright lines ; and of these the first discovered and by far the best known is the "1474 K" line, in the green, discovered by Prof. Young in the eclipse of 1869, August 7, and so called because he read its position—erroneously, as it now appears—as 1474 on the scale of Kirchhoff's spectroscope. We know as yet nothing of the gas indicated by this line, for it has not been discovered in any terrestrial substance; it has, however, been named provisionally, for the sake of convenience of reference, as " coronium." Eclipse work, therefore, at the present day usually includes in its programme the determination of the variation in relative intensity of these three sources of coronal light from one eclipse to another; the record and measurement of the positions of the bright coronal lines, and their identification, if possible, with terrestrial elements; and the distribution in the corona and round the sun, of the various bright lines of the coronal spectrum, and especially of the line " 1474 K," the line of " coronium."

From the first beginning of spectroscopic astronomy it was seen that the actual " photosphere," or bright surface of the sun, sent us light of every conceivable colour—in other words, would give us, if we could see it alone, a spectrum perfectly continuous, without dark line or break, from red to violet. The dark Fraünhofer lines were due to the absorptive action of gases closely surrounding the sun, and stopping out certain rays of definite colour or refrangibility. These gases being intensely heated, would, could we see them apart from the sun, give us discontinuous spectra of separate bright lines. Since, then, these gases, when the sun is viewed through them, cause the appearance of dark lines in its spectrum, they have been spoken of as " the reversing layer."

Up to the eclipse of 1870, December 22, the spectrum of these gases had not been separately seen. But on that occasion Professor C. A. Young, watching the spectrum of the dwindling arc of sunlight with a slit spectroscope, as the moon had all but hidden the sun, saw at the moment of second contact the ordinary solar spectrum with its dark lines on the continuous background disappear, and then " all at once, as suddenly as a bursting rocket shoots out its stars, the whole field of view was filled with bright lines, more numerous than one could count." This beautiful appearance has since become known as the " Flash."

The interpretation which Professor Young put upon this observation was that he had seen the spectrum of the " reversing layer." " Of course it would be very rash," he says, " on the strength of such a glimpse to assert with positiveness that these innumerable lines corresponded exactly with the dark lines of the spectrum which they replaced, but I feel pretty fairly confident that such was the case." This would mean that the dark lines which we see in the ordinary spectrum of the sun were due to the absorption of a comparatively thin shell of gases surrounding it, the spectrum of which we can only secure on such occasions as these.

It has therefore been a point much desired to secure a

photograph of the spectrum of the "flash," and this was actually done by Mr. Shackleton in the eclipse of 1896, and in 1898 the attempt was repeated and crowned with great success at nearly all the observing stations.

It was no exaggerated sense of its importance which, in recent eclipses, has placed photographs of the "flash" as absolutely the most instructive observations that could be made. For this narrow shell, passed over by the moon in about two seconds, and therefore about 800 miles in depth, is the lowest region of the sun's surroundings which we can study. If it be true, as Sir Norman Lockyer argues, that our terrestrial elements are broken up in the sun into bodies more truly elementary still, here we shall find the evidence of it. On the other hand, if there be a true "reversing layer," this is the region where we ought to find it.

Observations of the spectra which an eclipse can offer to us, and above all photographs of those spectra—for a photograph can record the position and intensity of a myriad of lines as easily as of one, and is not liable to make mistakes due to hurry or nervous excitement—are the most important which can be made. And of all the varied phenomena of an eclipse, the greatest interest at the present time attaches to the "flash."

The spectroscopes employed for this work are of many different kinds. Usually when a spectroscope is employed for astronomical work it is attached to a large telescope, taking the place of the eyepiece. The typical form of photographic spectroscope, or "spectrograph" as it is called for brevity, consists of a slit, placed in the focus of the great telescope, a collimating lens, to render the rays of light which have entered through the slit parallel, a prism, train of prisms, or diffraction grating, to disperse the light and produce the spectrum, and a photographic lens and camera to secure a negative of the spectrum when thus produced.

Such an arrangement would be termed an "analysing spectroscope," for an actual image of the eclipse would be formed on the slit plate of the instrument, and the light from such portions of the prominences and corona as lay across the slit opening, would be analysed by it, every point of light on the slit giving its own spectrum. But it is equally possible to use the spectroscope by itself, without attaching it to any large telescope, but simply pointing it up to the sky towards the eclipse. In this case no image is formed on the slit plate, but the general light from the whole phenomenon of the eclipse enters the slit together and gives rise to a single composite spectrum. A spectroscope so used is known as an "integrating spectroscope," and is of great value as a supplement to the analyser. Thus, for instance, in the eclipse of 1869, Young, observing with an analysing spectroscope, found the green line, commonly known as "1474 K," less bright than the greenish blue line near it due to hydrogen, and known as "Hβ," whereas Pickering, observing with an integrating spectroscope, found "1464 K" distinctly the brighter of the two. This showed that though "1474 K" was intrinsically much fainter than "Hβ," yet it existed over a much wider area than the hydrogen line. Less bright at one particular point, it gave a greater total amount of light on the whole, as it extended over a wider region. In short, "Hβ" was a line of the prominences, but "1474 K" was a line of the corona.

A third mode of using a spectroscope in an eclipse is dictated by the fact that as the black disc of the moon cuts out the light of the sun, the corona and prominences form as it were a rough ring of light, and a slit of a ring form is for many purposes just as useful in a spectroscope as the ordinary straight slit. If, then, when using the analysing spectroscope, i.e. the spectroscope in connection with a large telescope, we dispense with the slit, the eclipse itself will act as its own slit, and we shall have the spectrum not of a narrow strip alone, as we should if we used the slit, but of the entire phenomenon, every point of which would give its own special and peculiar spectrum. This method is usually briefly described as that of a "slitless spectroscope."

This method leads on to the fourth method. In this third method an image of the eclipse is formed at the point where the slit would have been, and the rays from that image are rendered parallel by a collimating lens before they fall on the prism. But we have already the actual eclipse itself up in the sky, and the rays proceeding from it are already parallel. We may discard, then, both the large telescope to form the image, and the collimator, and use only the prism and the camera. This brings our spectroscope down to its simplest form, and it is generally described as a "prismatic camera," or sometimes as the method of the "object-glass" or "objective prism."

One great advantage of this form is that it embraces the whole of the phenomenon in its grasp. Another is that it is very economical of light. A third is that its simplicity renders it possible to use a far larger prism and camera than could be done if these were but parts of a spectroscope attached to a larger telescope. Then it is the analysing spectroscope par excellence, for every part of the eclipse gives its own peculiar and separate spectrum; each separate coloured ray of light paints its own picture of the eclipse.

These four kinds of spectroscope were all in use by Mr. Evershed during the late eclipse. What may be described as a fifth form, a "prismatic opera glass," was used by Mr. Maunder. This consisted of a binocular, to one of the eyepieces of which a small direct-vision prism was attached. This enabled the observer to

compare the form of the corona as seen as a whole with its form as seen on one of the lines of the spectrum—say "1474 K,"—or in other words to compare the corona itself with the distribution in it of a certain gas, such as " coronium."

MR. J. EVERSHED'S REPORT.

THE work which I undertook at the recent eclipse was similar to that which I had intended to do in Norway in 1896, but was on a rather more extended scale. It consisted in obtaining spectrum photographs of the corona, prominences, and the reversing layer or flash spectrum.

On the former occasion I hoped to secure these photographs with one instrument only—a prismatic camera of $2\frac{1}{4}$ in. aperture and 36 in. focus (described in the *Memoirs* of this Association, vol. vi., part 1); but for the Indian eclipse I had, in addition to this instrument, a slit spectrograph containing two quartz prisms, intended for a single exposure on the corona during the whole of totality; and a large slitless spectrograph attached to a 6-in. telescope for photographing the flash spectrum on a larger scale than was possible with the prismatic camera.

Besides the three photographic instruments, I had available a 4 in. polar heliostat lent by Mr. Maw, and a $3\frac{1}{4}$-in. equatorial telescope, with solar spectroscope attached, for making visual observations on the bright lines, and to determine the exact moment when to expose the prismatic camera and large spectrograph in order to photograph the flash spectrum.

The heliostat was used to supply light to the prismatic camera and to the slit spectrograph. It was of the ordinary form, with two mirrors, but was modified for the special work by removing the second mirror and mounting it in the same plane as the first, so that two beams of light were available instead of one beam twice reflected.

Thanks to the facilities afforded us by the Indian Government in providing workmen and materials, and to the very attentive way in which all our needs were provided for by the Assistant Commissioner, Mr. D. O. Morris, Lieut. R.A., I was able to get all these instruments erected and in working order in time for the eclipse, although from the day of our arrival at Talni we had only a fortnight in which to make our preparations.

In setting up the instruments and in putting together the large spectrograph and 6-in. telescope I had also the advantage of receiving most efficient help from Captain P. B. Molesworth, R.E., without whose skilled assistance it would have been impossible to get all ready in time.

My observing hut was similar to those built for the other members of our party, the frame consisting of four uprights

connected together by cross-beams, and enclosing a space 12 ft.
square, in which the instruments were placed. The roof of the
hut was removable, and the sides, which were constructed of
light bamboo matting, could be easily adjusted so as to admit
or exclude light at any point.

The disposition of the instruments in the hut is shown in
the plan. The prismatic camera was screwed to two wooden
posts driven into the ground and bedded in cement. The long
axis of the instrument had to be carefully adjusted in azimuth

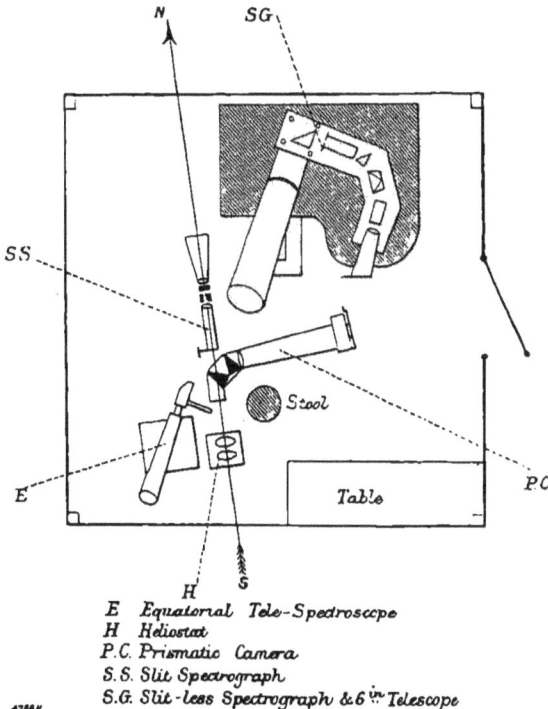

E   Equatorial Tele-Spectroscope
H   Heliostat
P.C. Prismatic Camera
S.S. Slit Spectrograph
S.G. Slit-less Spectrograph & 6 in. Telescope

GROUND PLAN OF MR. EVERSHED'S OBSERVING HUT.

and altitude, so that the refracting edges of the prisms should
be parallel to a line drawn tangent to the sun's edge at the
points of second and third contact. Otherwise the bright lines
of the flash spectrum would not be at right angles to the length
of the spectrum. A light metal exposing cap was fitted in front
of the prisms, which could be easily detached and replaced
without causing the slightest tremor.

The slit spectrograph was arranged in a north and south line,
with the collimator and image-forming lens directed towards
one of the mirrors of the heliostat. A blackened metal disc

attached by a hinge in front of the slit formed a very simple and convenient exposing shutter.

The slitless spectrograph was mounted on teak boards, which were bolted to the wooden tube of a 6-in. telescope, the collimator being placed at right angles to the axis of the telescope. A total reflection prism placed about 8 in. within the focus of the object-glass reflected the image of the sun into the collimator tube. The dispersion was obtained with a 60° dense flint prism * and two compound prisms of about $1\frac{1}{2}$ in. aperture.

The whole apparatus was mounted as low as possible on a rough equatorial stand having R.A. and Dec. slow motions, but no driving gear. The spectrograph end swung freely in a large pit excavated in the north-eastern portion of the observing hut, while the telescope was pointed directly at the sun.

When the telescope was directed to the place of the sun at mid-totality the object-glass and exposing cap came into a convenient position near to the other instruments, and about 5 ft. above the floor level.

The equatorial tele-spectroscope for visual observations was mounted on a pillar made by a packing-case sunk in the ground, and filled with earth and cement. It was placed to the south-west of the other instruments, and in such a position that when directed to the eclipsed sun the eyepiece of the small telescope was in a convenient position for the observer, who, seated near the prismatic camera, could easily manipulate the exposing caps and observe the spectrum at the same time.

As I had no one to assist me on the day of the eclipse, the three photographic instruments had to be arranged with their exposing shutters near together, so that I could work them all while seated near the tele-spectroscope.

The plan I proposed for starting the exposures at the right moment consisted in watching in the spectroscope for the actual appearance of the bright lines of the flash spectrum, then exposing simultaneously the prismatic camera and the large spectrograph. After that the slit spectrograph could be opened; then, after one or two short exposures, a long one could be started with the prismatic camera. During this exposure, which was to last 40 seconds, the large spectrograph was to be moved in R.A. to get into position for the flash spectrum at the end of totality, the plate-holder being at the same time reversed. After closing the 40-second exposure I was to expose one or two more plates in the prismatic camera for shorter intervals, and then close the slit spectroscope and prepare for the second flash spectrum; exposing the spectrograph and prismatic camera simultaneously as before, and closing the instant the photosphere appeared.

* Lent to me for this purpose by Mr. Otto Hilger.

On the day of the eclipse the actual procedure was as follows : About ten minutes before totality the heliostat was started going, and was carefully adjusted to get the spectrum central in the field of the prismatic camera.   Then the exposing cap was put on, and the first plate drawn up into position by the rack and pinion arrangement.   Then the exposing shutter of the slit spectroscope was closed, and the dark slide drawn out ready.

About two minutes before totality was due the large spectrograph was moved in R.A. until the image of the cusps touched a certain mark made on the slit plate exactly $\frac{1}{2}$ in. from the centre of the widely opened slit in the direction of the diurnal motion.*   This adjustment was easily made without assistance by looking at the slit image with a miniature telescope placed for the time being in front of the 6-in. object-glass, but facing the opposite way.   The slit being in the focus of the large O.G., the latter formed a collimator for the small telescope, in which the slit with the cusp image upon it could be clearly seen when focussed for parallel rays.   The eyepiece of the small telescope contained a total-reflection prism, so that I could look in at the side and not obstruct the incident light on the large lens.

The image of the crescent sun was kept on the mark by following in R.A. until the chronometer I was using indicated 88 seconds before totality, then it was allowed to drift; I then quickly removed the small telescope, covered up the 6-in. O.G. with a plate of blackened aluminium, and drew out the camera dark slide.

During the last half-minute before totality was due I began exposures with the prismatic camera, taking two instantaneous photographs of the cusp spectrum and then drawing another plate into position ready for the flash spectrum.

Now, all being ready, only a few seconds remained before the bright lines might be expected to appear.   The gloom of the approaching shadow was already increasing at an alarming rate. I turned to the tele-spectroscope, took off the slit head, and watched the spectrum of the last remaining thread of sunlight without any slit, observing in the green region near b.   All the curved dark lines of the ordinary spectrum were seen at first just as though the semicircular slit ordinarily used for observing prominences had not been removed.   But almost immediately the rapidly narrowing band of continuous spectrum broke into a number of strips, and the dark lines disappeared ; at the same moment the bright lines flashed out in hundreds between and across the streaks of continuous spectrum.   They were not faint, short lines, but long and brilliant arcs very sharply defined and extending in many cases over 30° of the limb.

The most astonishing part of this beautiful display was the

* The large spectrograph was fitted with an ordinary slit, but, as the jaws were opened very widely, it was practically a "slitless" spectrograph.

instantaneous transformation of the lines from dark to bright the moment the continuous spectrum broke up.

I did not wait for any further developments but immediately uncovered the prismatic camera and the large spectrograph, giving each instrument an exposure of several seconds before replacing the caps. A few seconds after second contact I made an instantaneous exposure with the prismatic camera, and then started a long exposure, at the same time opening the slit spectrograph. Then I left the seat and went to the camera of the large spectrograph, closing the slide, reversing, and opening again ; then the R.A. handle was turned four revolutions, to bring the whole instrument again into position for the flash spectrum, which at the end of totality would occur on the western limb.

Returning to the seat, I closed the prismatic camera long exposure, and started another. During this exposure I took a hurried look at the corona with a pair of fieldglasses. I saw the planet Venus, and noted the shape of the great S.W. streamer, but had no time to look for the prominences or other details : 20 seconds only, according to the time-caller, remained before the sun would reappear. It was therefore necessary to prepare for the second " flash." I exposed the prismatic camera once more—the last in totality—and then, at the moment when I was conscious of returning light, again made a simultaneous exposure with the prismatic camera and large spectrograph.

After that two snap-shots with the prismatic camera out of totality, following as rapidly as possible, completed the programme.

The whole performance seemed to have gone off without any hitch ; but too late I discovered the slit spectroscope still exposed, with the crescent sun right across the slit ! I had forgotten to close the shutter in the hurry of the last moments of totality.

It will naturally be inferred from this account that I attempted too much and had too many instruments for one person to attend to. It would have been far better, no doubt, to have had one assistant for each instrument, and one besides to record the times of exposures. Unfortunately the assistants were not forthcoming. I am confident, however, that with a day or two devoted to drilling there would have been no difficulty whatever in going through the operations I have described without any mistake. As it was, owing to the unavoidable delay in getting into camp, the work of erection and adjustment occupied the whole time even up to the hour of the eclipse, and what little drilling I obtained had to be done in a few spare moments. What I most regret is, that I was unable to obtain a really satisfactory view of the eclipse itself.

The number of photographs secured altogether was thirteen : one with the slit spectrograph, two with the slitless spectrograph, and a series of ten with the prismatic camera. The single photograph obtained with the slit spectrograph failed from the above-mentioned cause, the direct sunlight and halation nearly obliterating the faint corona spectrum. The slitless spectrograph yielded two negatives of the flash spectrum at second and third contacts. They show a large number of bright lines in the region between F and H, but on the whole they do not quite come up to expectation. The most interesting results were those obtained with the prismatic camera. This instrument gave images of the spectrum extending from λ 6000 in the orange to λ 3350 in the ultra-violet and on a scale of ·33 inch to the moon's diameter, the length of the spectrum between the above limits being nearly three inches.

All the ten photographs of the series yielded interesting results; the flash spectrum appears on Nos. 3, 7 and 8, while the long exposures about mid-totality give the corona spectrum, and those taken out of totality show the Fraünhofer dark line spectrum bordered with bright lines.

In No. 3 the flash spectrum lines are beautifully defined in the ultra-violet, where they can be traced as far as λ 3342. In this region, from H upwards, 218 lines can be counted. Through the kindness of Dr. Rambaut, who supplied me with an excellent measuring apparatus, I have been able to determine the wave-length of every line with a very satisfactory degree of accuracy.

From the results obtained it appears that a great many of the lines are due to iron, some to calcium, magnesium, etc., and three of the strongest lines in the spectrum have been identified by Mr. Jewell (at the Johns Hopkins University) as due to the comparatively rare element titanium. The vapour of this metal is not, however, confined to the " flash " layer, but extends as high in the chromosphere as hydrogen, and enters into the composition of the prominences as well.

In this plate (No. 3), there are thirty hydrogen lines shown ; they follow each other with mathematical regularity as indicated by Balmer's well-known formula, and with gradually diminishing intensity. The Greek alphabet is, of course, exhausted in designating them, starting with the C line as a, in accordance with the usual procedure. The wave-lengths actually obtained for these lines agree remarkably accurately with those derived from the formula. A very slight deviation from exact agreement occurs, however, in the lines beyond the ω line, but these, on account of their faintness, are subject to a greater uncertainty than the others.

In the visible portion of the spectrum (H to D), it is not possible to determine the wave-lengths with anything like the accuracy obtained in the ultra-violet ; but for the sake of completeness, every line on the photograph has been measured and its wave-length determined. The result obtained for the

No. 1.—Exposed 20 Seconds before Second Contact.

ECLIPSE SPECTRA.

From photographs taken with a prismatic camera, 2¼ inches aperture, 36 inches focus.

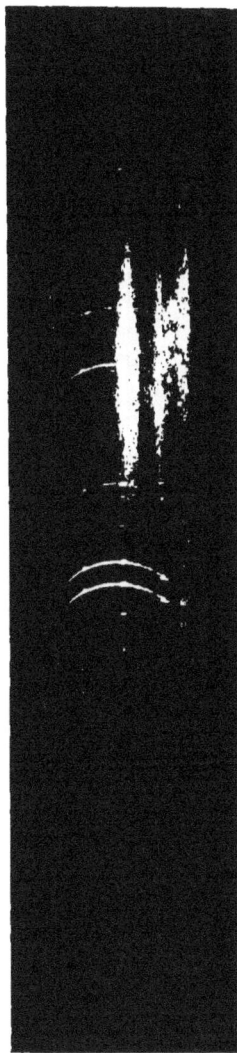

No. 4.—Exposed 10 Seconds after Second Contact.

ECLIPSE SPECTRA.

From photographs taken with a prismatic camera, 2¼ inches aperture, 36 inches focus.

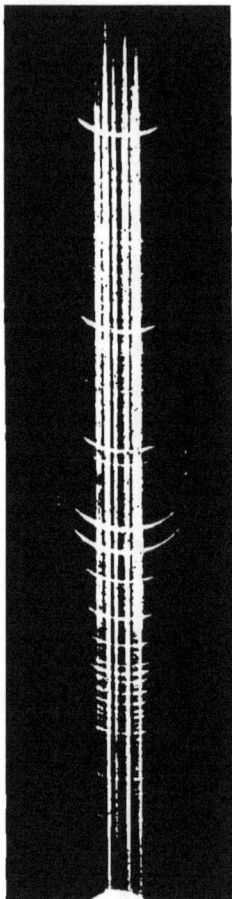

No. 8.—FLASH SPECTRUM AT THIRD CONTACT.

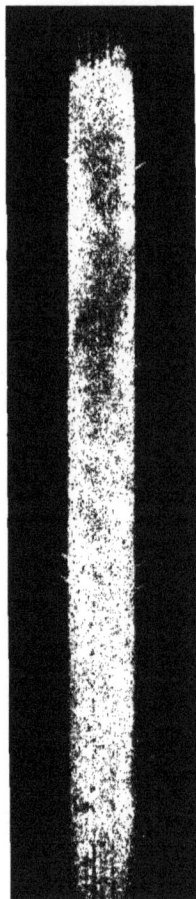

No. 9.—EXPOSED 10 SECONDS AFTER THIRD CONTACT.

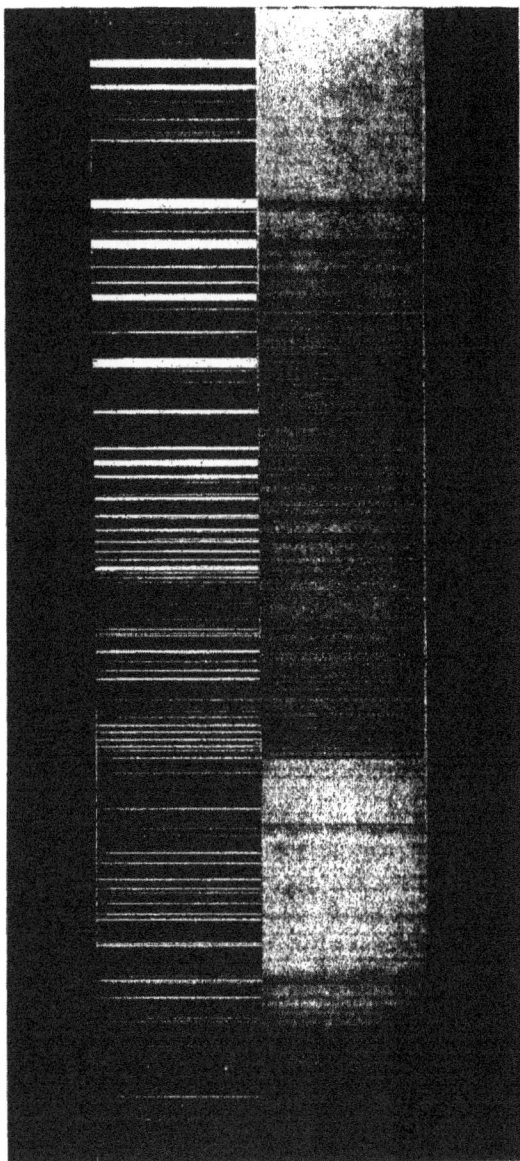

ECLIPSE SPECTRA.

From photographs taken with a prismatic camera, 2¼ inches aperture, 36 inches focus.

FLASH AND CUSP SPECTRA COMPARED. ULTRA-VIOLET REGION λ 4100 TO λ 3350.

ENLARGED FOUR TIMES FROM NEGATIVES NOS. 3 AND 1.

corona line ("1474 K") is interesting, as it entirely confirms the wave length obtained by Mr. Fowler for this line. The mean of measures of the line on Plates Nos. 3 and 7, gives a position about 14 units more refrangible than the value (λ 5317) which has for so long been accepted as the true value.

Six of the ten photographs of the prismatic camera series are reproduced in the accompanying plates, the series number and time of exposure being given under each spectrum. They are enlarged from the original negatives about twice, and show the principal features very well. The hydrogen lines are well seen in Nos. 3 and 7, the ultra-violet members of the series to the left hand of the pair of strong lines H and K, which will be easily recognised on each spectrum near the middle of the plate. The first strong line to the left of K is the hydrogen line ζ, and the others follow at regularly decreasing intervals, and diminishing intensity. It will be observed, however, that there is a strong line near the end of the series (between ρ and σ), and a close pair of equally strong and long lines between ι and κ, which appear to spoil the harmony of the series : these are the three titanium lines referred to above.

With a view to making a direct comparison between the flash spectrum and the Fraünhofer spectrum, I have prepared a plate in which these two spectra are placed side by side. By photographing a narrow longitudinal strip of No. 1 and No. 3 spectra with the interposition of a cylindrical lens, the short sections of the curved arcs are spread out into straight lines, which give to the spectra the conventional appearance, and aid in the comparison.

The Fraünhofer lines in No. 1, although much weaker than in the ordinary solar spectrum, are nevertheless very clearly defined in the original negative, and a comparison can be made with very great exactness. From a mere inspection of the plate in which these spectra are compared, it will be noticed that whilst some of the strong lines in the flash are reversals of strong Fraünhofer lines, a large number of conspicuous flash lines appear to have no corresponding dark lines. The calcium lines H and K, and many iron lines, are instances of the former, whilst the titanium and ultra-violet hydrogen lines may be taken to represent the latter. Many of the strong flash spectrum lines are, however, in reality represented by dark lines in the Fraünhofer spectrum, only the relative intensities are very different in the two spectra. Thus the titanium lines referred to above have their counterpart in the Fraünhofer spectrum, appearing in the ordinary solar spectrum as fine lines, which are, however, too weak to show at all in the cusp spectrum.

It would be out of place in this report to describe in any further detail the results which follow from a study of these photographs, or the bearing of these results on solar theory. It is desirable, however, in view of the approaching Spanish eclipse, to give an outline of what I consider to be the most important subjects for future investigation, and the most hopeful methods of improving upon the work done at the recent eclipse, and I have accordingly done so in a separate article on page 154.

Mr. E. Walter Maunder's Report.

My instrument was a binocular of about two inches aperture, in one of the eyepieces of which a small direct-vision prism had been mounted. The instrument was so arranged that both halves of it could be used at one time, and so the actual image of the corona could be compared with the ring of light given by any particular bright line in its spectrum. My purpose was to compare the so-called " 1474 K " ring with the corona so as to ascertain the distribution in it of the element " coronium." Each tube of the binocular was furnished with a pair of neutral dark glasses.

I did not commence any regular observations with the binocular until the partial phase was far advanced, though I looked at the spectrum from time to time.

By 19^h 15^m, Greenwich Mean Time, a vague suggestion of curvature had come into the spectrum—a convexity towards the red. The spectrum was arranged to lie along the line of the moon's motion, the red lying in the direction towards which the moon was moving. About 19^h 30^m the D lines were distinctly seen as a faint, diffused, crescent-shaped shading, and a second similar one was seen in the green, probably the b lines. These gradually became more distinct, and by 19^h 45^m C and F were also seen. From this time the Fraünhofer lines began quickly to multiply and to become more and more distinct, until at length, some ten minutes before totality, the spectrum was crowded with semicircular arcs, which grew finer and sharper as the arc of sunlight narrowed. The definition of the arcs was very fine, the focus being perfect.

Then the tips of the arcs turned to light as the continuous spectrum began to narrow. C, D_3, F, and G, were particularly noticed.

The changes now came very fast, the spectrum narrowed, and more bright arcs of light appeared. The spectrum did not however narrow uniformly, so as to make the central part disappear last. Just before it was finally lost, it broke up into a number of very fine threads of continuous spectrum. At about the same instant the spectrum was full of an infinite number of bright points, beside the arcs already seen. The

PROF. NAEGAMVALA'S CAMP, SHOWING HORIZONTAL TELESCOPE AND CŒLOSTATIC MIRROR.

threads of continuous spectrum snapped and disappeared, and totality had begun; and I gave the signal "Go," and at the instant the bright points disappeared also.

I had experienced a great difficulty in the observation of the last minutes of the partial phase. The sun was intensely bright, and both dark glasses were none too much for the sun itself, and one dark glass was rather too little for the spectrum. But it became quite clear to me that it was risky to leave changing all the dark glasses until the very moment of totality, as the changes then would obviously be very rapid. I therefore discarded one of the dark glasses from the direct image quite two minutes before the total phase began, and one minute later dropped the dark glass from the spectrum. In the next few seconds I discarded the last dark glass from the direct image, but kept that eye, the left eye, shut until after totality had begun, observing only on the spectrum with the right.

The spectrum even in the last minute was painfully bright (I had tried again and again earlier to see if it were possible to observe without the one dark glass on the spectrum, or without the two glasses on the actual sun, but had found it impossible). When therefore totality began I fancy my eyes were anything but sensitive. But for a long time I could detect no trace of the "1474 K" line.

The spectrum of the corona was in the main a bright continuous spectrum of irregular brilliancy. The C, D₃, F, and G lines were seen as brilliant sharply defined semicircles, and several very much shorter arcs were seen for a little time, the b triplet being one. But failing to detect "1474 K," I turned my attention after a time to the corona itself.

When I turned back to the spectrum, the coronal line was clearly seen, though very faint. Mid-totality was then past, and the line was only seen on the west side of the moon. It was traced through something less than half a circumference and more than one-third—say 140°.

It was much fainter than the arcs of the C, D₃, F, and G lines, but was broader, perhaps twice as broad. It was also diffused on its outer edge, not sharp as the other four lines were. But I could trace in it neither rifts nor rays. I estimated its breadth (at the cry of "twenty") as between $\frac{1}{5}$ and $\frac{1}{8}$ of the moon's diameter, or about 6′ of arc.

The eclipse now drew rapidly to its close; at the cry of "ten" the C, D₃, F, and G lines were evidently brightening, and other shorter arcs were appearing. In the next four seconds the bright lines multiplied exceedingly, a number of bright threads of continuous spectrum flashed out and ran together, a flood of sunshine broke out and the eclipse was over.

The time-keeper, Mr. Ramrao Subarao, clerk to Mr. Morris, called out "Close cameras" at the fifth beat after the ten seconds, but the eclipse was then over by a full second. As the clock had been arranged for 118 seconds, the eclipse had lasted 112 seconds, or 4 seconds less than had been expected.

It has long been known that the coronal spectrum is from three distinct sources. One, a purely continuous spectrum, probably from glowing particles of cosmical dust ; a second, with difficulty distinguished from it, of reflected sunlight, *i.e.* a continuous spectrum interrupted by the dark Fraünhofer lines ; and the third a bright-line spectrum due to glowing gases. It would seem, from the comparison of the observations of different eclipses, that the third kind of spectrum varies in its brightness, relatively to the sum of the other two, very much from time to time. The coronal spectrum in 1893 and 1882 and 1883—*i.e.*, years of maximum—seems to have given readier evidence of bright lines than in years nearer minimum. The present occasion was therefore not a good one for such an observation as I had undertaken ; and if repeated in 1900, a much higher dispersion should be employed in order to get rid as far as possible of the light from the first two sources.

So far as my observation at this eclipse goes, it points to the gas, which we may call "coronium" for distinctness, being distributed pretty evenly round the sun to a distance of about 160,000 miles, and that it is not specially associated with the striking irregularities of the visual corona. It must be however remembered that the corona as seen is far brighter up to about this limit than above it ; that the "inner corona," in short, corresponds pretty closely to this distribution of "coronium."

MISS DIXON'S REPORT.

MISS DIXON, assisted by Miss Bevan as recorder, observed at Buxar with a similar prismatic opera-glass, kindly lent by Miss E. I. Page ; and noticed that at six minutes before totality the Fraünhofer lines were clearly seen, particularly C, D, F, and G (C and G being specially marked). These were similarly seen after totality, and lasted fully ten minutes. Their first disappearance was immediately before totality. The phenomena of the reversing layer was clearly seen at the commencement and end of totality, many bright arcs flashing out, C, D_3, b, and F being very conspicuous. The bright arc corresponding to "1474 K" was specially looked for, but could not be identified.

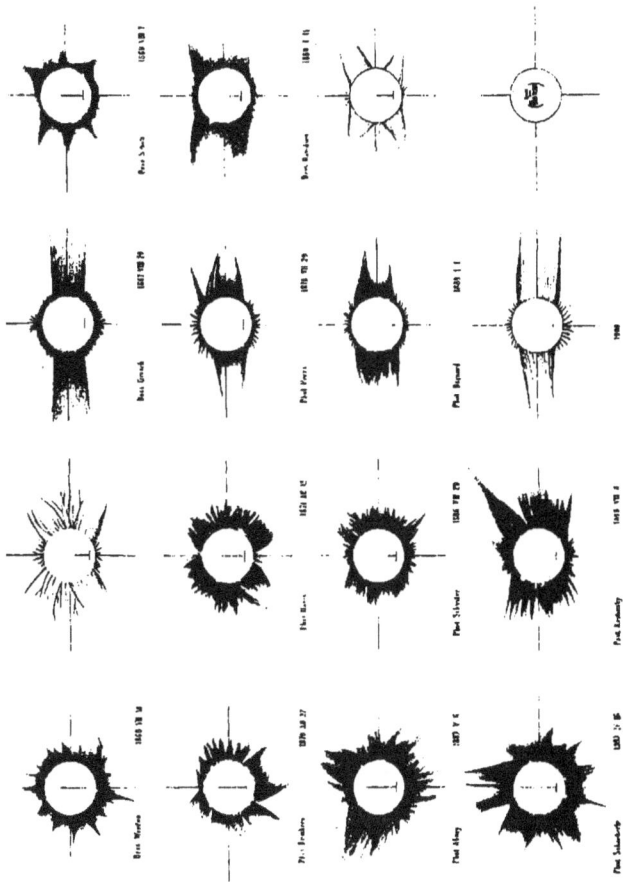

At Maximum.   After Maximum.   At Minimum.   After Minimum.

FORMS OF CORONÆ AT DIFFERENT EPOCHS.

("Dess." signifies drawing; "Phot." photograph.)

CHAPTER VI.

THE APPEARANCE OF THE CORONA.

WE cannot but regard it as a strange circumstance that there is so seldom any mention of the Corona in the records of ancient eclipses. It is often mentioned that "the stars appeared as in the night-time," though we know from the experiences of recent years that few stars other than the brightest planets ever become visible on such occasions. But the corona, so beautiful an object and of such · strange and mysterious shape, larger than the moon, and often much brighter than the moon when full, is scarcely ever unmistakably alluded to. Plutarch is generally held to have referred to it in a well-known passage describing a total eclipse, and Philostratus in his life of Apollonius of Tyana; whilst a "red light" which was seen round the dark moon, during the solar eclipse which took place during the battle of Sticklastad in Norway, in 1030, August 31, is generally thought to have been the corona, but may have been a brilliant display of prominences. In more modern times we have a few explicit and definite accounts, ranging from the observations of Clavius, 1567, April 9, down to those of the eclipse of 1842, July 8, from which time the corona has been the subject of careful and attentive study.

The explanation of this strange omission of the corona from nearly all the old accounts, whilst the appearance of the stars is so frequently noticed, is probably that the old observers knew that the stars ought to appear, and therefore looked for them, and in consequence saw them. But they did not expect to see the corona, and if they did notice it very likely thought it a mere diffusion of the sun's light from behind the moon, just as we see the sun's rays lighting up the dusty or moisture-laden air when the sun itself is hidden behind a cloud, whether due to the effect of an atmosphere round the moon, or to a scattering in the higher regions of the earth's atmosphere, or to a sort of diffraction effect at the moon's limb. Each of these three explanations has been widely accepted in its turn; that which ascribes the corona to the scattering of the sun's light by our own atmosphere being strongly held by some even as late as 1870.

It was not altogether unnatural that there was a reluctance to accept the corona as truly solar until it had been proved to be such by the clearest demonstration. For if it belonged to the sun, its extent must be counted even by millions of miles, and it must form a structure of a vastness beyond the power of man's imagination to truly appreciate. Yet the proof that it was solar had been supplied as early as 1724 by Maraldi, who observed that the corona did not travel with the moon but was traversed by it; but this deduction found no general acceptance until the eclipse of 1851, and some astronomers held out against it for two decades longer.

The earliest drawings made of the corona are very curious. They give it the form of a Greek cross—four equal arms at right angles to each other. Two drawings preserved to us in 1715, and a third in 1766, and reproduced in the great eclipse volume published by the Royal Astronomical Society, vol. 41 of their Memoirs, well illustrate this peculiarity. Later we find the tendency of the drawings to run into general halos, the desire of the artists evidently having been rather to suggest the general effect than to delineate precisely the form and structure of the corona as they actually saw it.

From 1851 onwards we have a large number of drawings which were evidently made with due care and conscientiousness. They are not always easy to collate together, because, as eyes differ in sensitiveness and as atmospheric conditions differ from place to place, so the form and extent of the corona as drawn by different observers in the same eclipse vary exceedingly. A striking instance of this was afforded in the eclipse of 1874, April 16, when two observers, Mr. Henry Hall and Miss Alice Hall, seated side by side at the same table, drew the corona with results which did not bear the slightest resemblance the one to the other, Mr. Hall tracing the outline of the corona to not quite eleven minutes from the moon's limb, whilst his sister traced it in one direction nearly ten times as far. Fortunately before the eclipse was over they were able to compare notes, and it became clear that whilst Mr. Hall had drawn the outlines of the brightest and innermost portion of the corona, his sister had endeavoured to depict the shape of its faint outer extensions.

The eclipse of 1870 was well observed in Spain and Sicily, and that of 1871 in India and elsewhere, and a valuable harvest of drawings and photographs was secured. The next eclipse that was observed with anything like the same fulness fell in 1878, the track then lying across the American continent and calling forth a great army of skilled observers. It was at once seen that the corona of 1878 was utterly unlike in form those of 1870 or 1871. In the two earlier years it had had the form of a great irregular halo, sending out its rays indiscriminately in almost every direction round the sun. In 1878 there was a

Mr. Nicolson.   Mrs. Pevan.   Mrs. Oakes.   Mr. Moore.   Mr. Johnson.

THE SKETCHING PARTY AT BUXAR.

comparatively narrow bright ring of light round the sun, flanked by two enormous extensions east and west in the plane of the sun's equator. It was impossible to overlook so striking a change ; and on comparing the drawings of 1878 with one which had been made eleven years previously in the eclipse of 1867 by Grosch, it was seen that the two coronas were almost of exactly the same shape. This striking change was at once connected with the change which had taken place in the same time in the sun's surface. 1870 and 1871 were years in which the spots upon the sun were unusually numerous and large ; 1868 and 1878, on the other hand, were years when the sun was almost entirely free from spots. Looking back to the eclipse of 1860, which was very fully observed, chiefly in Spain, and of which Mr. Weedon's drawing in the plate on p. 86 may be taken as a specimen, the corona was seen to be utterly unlike those of 1867 and 1868, and to closely conform in character to those of 1870 and 1871 ; and it was noted that 1860, like these two last-mentioned years, was one of many sun spots. From that time forward a correspondence between the general form of the corona and the development of spots upon the sun has always been looked for. The plate on p. 86, which is reproduced from M. Backlund's Report on the Russian Expedition to Novaya Zemlya, 1896, represents the general form of the corona in the eclipses from 1860 onward. The first column shows the coronæ of 1860, 1870, 1883 and 1893, at all of which dates the sun was very largely spotted—sun spots were at their maximum. The second column shows the coronæ in the years 1871, 1886, and 1896, when the solar activity was declining. It would have been a more instructive comparison had the corona of 1874 been substituted for that of 1871, as the decline had scarcely commenced in the earlier year. The third column gives the years of minimum of 1867, 1878, and 1889, with a forecast for 1900. The last column represents a phase for which as yet we have not much material—the period of increasing solar activity. It will be noted that the first three types are very distinctly defined.

Mr. H. Keatley Moore's Report.

In the British Astronomical Association's Eclipse Expedition to Norway, 1896, our late President (Mr. N. E. Green) organised a section for drawing the corona as seen by the unaided eye. Twelve members placed themselves at his disposal, and one or two meetings for practical preparation were held. Mr. Green decided, after mature consideration, to draw with white chalk upon purplish-blue paper, as best representing with simple and easily managed materials the conjectured appearance of the white corona upon the darkened sky : and as it would be difficult for a trained artist, and impossible for an amateur, to sketch accurately the whole of a complicated object like the corona in the minute and a half, or thereabouts, which is all the time allowed by nature, Mr. Green subdivided his section into four groups of three draughtsmen each, and allotted one quadrant to each group, settling at the same time the size of the circle representing the eclipsing moon, so that all the drawings would be upon the same scale. White chalk, carefully pointed, allows the artist to draw lines of varying force and thickness with the point, and also to express broad sweeps of radiance rapidly by using the side; it is therefore admirably adapted for simple rough work of this kind.

As soon as the whole of the second division of the India Expedition had gathered on the *Egypt*, the work was divided out at a general meeting, and it was found that four persons accustomed to sketch from nature were willing to constitute such a section as Mr. Green had conceived. These were Mrs. Bevan, Mrs. Oakes, Mr. Nicholson, and Mr. H. Keatley Moore. Mr. Moore was chosen to direct the party, and it was agreed to follow as closely as possible the lines laid down by Mr. Green. Blue paper not being procurable in mid-ocean, brown paper was necessarily substituted at the practices, but the actual drawings were made upon paper of the desired tint. The enthusiasm of the little section enabled practices to be held for about an hour nearly every evening after it was constituted. The observers sat in a large space on the lower deck, beneath the great well-opening of the after hatch. Each evening an imaginary corona was drawn to a large scale, roughly resembling one of the coronas of previous eclipses : the drawing was pinned to a stiff board in which holes were bored along each edge, so that it could be hung up by any of its four edges, and then this copy, adjusted at such a height up the hatchway as corresponded to the height of the sun in the sky on eclipse day, was illuminated by a very powerful electric lamp which Capt. Briscoe kindly allowed the electrical engineer (Mr. Malden) to arrange nightly for this purpose. Some one kindly stood upon an upper deck at the opening into the hatchway ready to expose the copy at a signal, and to withdraw it after ninety seconds : and another volunteer kindly called the time every five seconds by a watch. It was therefore as nearly as possible practising under actual eclipse conditions.

Each observer was provided with a plumb-line, so that the imaginary moon and corona were cut vertically; and as the horizontal division was not difficult to guess, the four quadrants were thus divided alike by all. A half-crown served to draw the moon-circle for the draughtsmen (the moon of the copy being usually 3 inches in diameter), and the sketching paper was divided into four parts by two chalk lines crossing at right angles at the centre of the moon-circle. [No such divisions were shown in the copy.] Mrs. Oakes always took the top right-hand quadrant, Mrs. Bevan the top left-hand ; Mr. Nicolson the bottom right-hand quadrant, Mr. Moore the bottom left-hand. After 5 seconds' warning the copy was exposed, the counting began, and each drew his own quadrant until at 90 seconds the copy was withdrawn. All the drawings were then collected on to one paper, by being copied, and the result was critically compared with the original. The original was then rotated, and suspended by another side, so that each observer had a fresh quadrant to draw. After four drawings each observer thus had a complete corona drawn by himself ; and there were also four combined drawings in which the errors of one observer were partially annulled by the work of the others. It soon appeared that the observers had a "personal equation," and that while some were always too timid, others were always too bold, etc. But after a few nights of steady practice all drew closely together, nervousness disappeared, and the combined results became excellent. All measurements and comparisons were invariably made in terms of the diameter of the eclipsing moon.

On eclipse-day the danger was, of course, the fatigue to the eye induced by succumbing to the temptation to look too often at the sun during partial eclipse, either through smoked glass or through some of the excellent telescopes. (The Rev. Septimus Hebert's capital instrument proved an especial snare in this respect.) The director of the section was perhaps over-eager in guarding his flock. For a few minutes before the eclipse became total the whole four sat in position (as arranged at rehearsal on the previous day) perfectly prepared, with plumb-lines adjusted in front of their chairs, etc., but with closed eyes ; and they did not look up until by the beginning of the counting they knew that the corona was visible. In spite of all these precautions it was found that one of the party had used his eyes too freely beforehand, and his quadrant suffered thereby, but the practices had been of such service that the observers of the adjoining quadrants were able to supplement the deficiency, having had time to go beyond the borders of their own simpler quadrants. The knowledge that Venus was 12 diameters away afforded an invaluable measure.

The whole of the afternoon was spent in making a most careful combined drawing ; other observers kindly aiding the section with criticism, which was freely welcomed. The result, when agreed upon, was at once photographed for greater security, as well as the separate drawings. Unhappily the plate had already served ! Such accidents will occur. The original chalk drawings, however, although of such fugitive material, were very carefully protected, and are still uninjured. They have, of course, never been touched since.

The reproduction of the composite drawing, given on p. 90, should be turned so as to bring the word "vertex" to the top, in order to view the corona as seen.

Three of the four sketchers (all used to sketch in colours) agreed that the corona was very nearly white in colour, with a faintly amethystine bluish tinge such as an arc light often gives. The fourth observer is the one referred to above as having fatigued his eyes. He considered the light to be pure white.

Two other members of the Association, Mr. T. W. Backhouse and Mr. J. Willoughby Meares, were present at Buxar, and made outline sketches.

MR. BACKHOUSE'S REPORT.

FOR a short time before totality I covered my eyes with my hands, keeping them mostly shut. I did not notice either the time or the duration of the totality. A gentleman says he found it 1 minute 32 seconds. I observed the totality mostly with field-glasses. I directed my attention specially to the outer parts of the corona. The figure shows as much as I could distinctly notice with field-glasses in the short time. I could not see any alteration in it during totality, except, I believe, a slight increase in the apparent lengths of the rays, due evidently to increasing power to detect them through continued observation, rather than to increasing sensitiveness of the eye to faint light. The main features were, however, extremely conspicuous and very striking. There were several rays—five more conspicuous than the rest—and to these five alone I paid special attention. A was the most conspicuous; next B and C; next, I think, D. Either D or E was very much narrower than the others, and not nearly so tapering; I believe this applies to both, especially E, but am not sure as to this. The narrow one, whether D or E, was perhaps intrinsically as bright as any, but less conspicuous because of its narrowness. All were colourless as far as I could judge. I should call them pearly white, at any rate the three chief. I could see no striations nor structure of any kind in them. Their outline was almost quite definite; they did not fade away gradually at the edges nor materially at the points, though their brightest portion was certainly that nearest the moon. I carefully estimated their lengths in diameters of the moon, as follows :—A, $2\frac{3}{4}$; B and C, $1\frac{3}{4}$; D, $1\frac{1}{2}$. I am not sure that I have drawn the widths at the base correctly. I looked very carefully for anything beyond these rays, both

with field-glasses and spectacles, and with my naked eye, but
could not see the least trace. Indeed the sky looked very
uniform in colour and darkness for some distance off. It was
blue, a very strong blue, yet not exactly the colour it has in ordinary daylight. I thought there was a purplish tinge, especially
just after the end of totality, at which time for some seconds it
seemed to me there was a very strong red tinge in the sky at
no great distance from the sun ; it might perhaps be the reddish
circle that appears round the sun on dusty air, and which has
often been visible all day since we came to India, though more
often in the middle of the day there has been merely a glare
without any red tinge. The red after totality seemed much
too strong to be due entirely to this, and struck me as probably
at least partly due to the light of the chromosphere. During
totality I could not see the slightest effect of dust nor anything

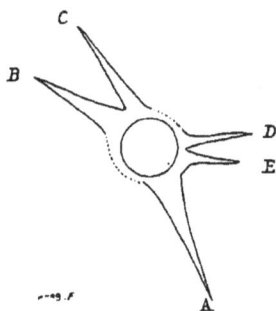

MR. BACKHOUSE'S SKETCH.

to hide the faintest parts of the corona beyond the general light
of a clear sky.

As regards the inner part of the corona, it formed a very
brilliant circle, far more brilliant than the rays. It rapidly
faded away outwards, but I did not notice how far it extended.

The chromosphere was very conspicuous with field-glasses,
especially just before the end of totality, when it was chiefly at
the right side of the moon. It was perhaps equally conspicuous
at the beginning of totality ; but I paid no particular attention
to it at any time. It was a brilliant crimson forming a rather
irregular rim round the moon, perhaps not quite continuous,
though nearly so. My impression is that it was visible all
through totality, but I am not sure. I did not see any *very*
lofty prominences.

The moment the sun began to reappear, the rays of the
corona disappeared ; but I think its inner part did not do so at
once. I also looked for Venus with the naked eye a very few
seconds after the sun appeared, and failed to see it.

MR. WILLOUGHBY MEARES' REPORT.

MR. J. WILLOUGHBY MEARES watched the progress of the partial phase by projection from a pinhole on a glazed cuff. When about half obscured all the lights thrown on the ground and the tents, through chinks in the trees, became markedly of the shape of the uneclipsed portion.

At about 13^h 25^m the colour of the sky had become an intense neutral tint, the air had cooled very perceptibly, and the kites had become very excited.

Totality commenced at about 13^h 43^m (Madras time). No

Venus about 10 diams.
(west)

MR. WILLOUGHBY MEARES' SKETCH.

shadow was observed crossing the country, either where we were or from the higher ground at the fort where a number of sight-seers went specially to see this.

The long *coronal streamer* on the lower side became visible some 2 seconds before totality and continued visible for some 5 seconds after totality had ended. For the last second before totality the uneclipsed portion appeared like a brilliant star, and not elongated at all.

No sketches were made at the camp at Talni, but the following descriptions of the corona have been supplied :—

MR. J. P. G. SMITH'S REPORT.

THE first contact was at 11^h 45^m 2^s local time, obtained by comparison of chronometer with the signal by telegraph at Talni from Madras observatory at 4 p.m., 21st inst. The gradual diminution of the light was distinctly sensible, and animals and birds showed that they felt this by returning homewards as in the evenings.

I did not notice the "stalking towards us" of the shadow,

but a sensible impression of green colour was made when about 0·8 of the sun's disc was obscured. The crescent shape of this phase was seen in the multitude of pictures of the sun, cast upon the floor of the observing tent, formed by the light penetrating the interstices of the bamboo matting that formed its southern wall. At the instant of the sun's total obscuration the outburst of the glory of rays constituting the corona was a sight never to be forgotten : it was a fringe of light extending generally about a whole diameter of the sun, say half a degree, beyond the darkened disc ; but this fringe was much elongated in a fishtail shape on the N.E., S.E., S.W. and N.W. sides of the sun, that on the S.W. or right lower side of the sun being very distinctly more elongated than the other portions. I estimated this to extend as much as two diameters, or *perhaps* a little more than $1\frac{1}{4}°$ from the sun, while the elongations of the three others varied from three-quarters of a degree to one degree.

MR. WALTER MAUNDER'S REPORT.

MY first impression was that the corona was, roughly speaking, rectangular in the general outline of its brightest part, the upper edge being practically parallel to the horizon. I judged the horizontal sides of the rectangle were to the vertical as perhaps 5 to 4.

The next point that impressed itself on me was the presence of four streamers, noticed in the following order.

I. N.E. A fine bright streamer traced quite $1\frac{1}{2}$ diameters from the moon's limb.

II. S.W. A longer streamer, but perhaps not quite so bright as I. It was traced $2\frac{1}{2}$ diameters from the moon's limb. I judged it to be nearly in the sun's equator, but slightly to the south of it.

III. N.W. A much shorter streamer, perhaps $\frac{3}{4}$ diameter.

IV. S.E. A fainter streamer. I just recognised it, but no more, and paid no special attention to it.

The third point noticed was the presence in the first three streamers of the typical coronal curves. These were especially noteworthy in the N.E. streamer, *i.e.* in its lower portions. The bulb of a flask or of a hyacinth, or the curve of a leaf, suggested themselves as like the outlines here shown. All three streamers, I., II., III., showed them ; they seemed most pronounced on I. ; most definite, most bright, and the curves the strongest.

Several prominences were noticed, one at the S.E. point especially of a full garnet hue ; two much smaller N.E., and at the last moment three on the N.W. limb.

To my eye the corona appeared of a silvery white, or better, a French grey with a lustre as of polished silver.

The general appearance of the corona strongly resembled that of 1886 as I remember it, and, judging from prints and photographs, those of 1868 and 1896 ; but it seemed to me much brighter and larger than that of 1886, though in shape it was very like it.

The eclipse was a very light one. The darkness was not to be compared to that of 1886 or even 1896. The general skylight was very considerable. I saw no other star than Venus.

After the eclipse was over, the gradual return of *colour* to the landscape was very striking. During the eclipse, except for a sunset yellow or amber light in the sky near the horizon, the whole looked like a landscape viewed through a purple-tinted glass.

It may not be out of place to add here a description of the eclipse which, though not contributed by a member of the Association, has a very deep and melancholy interest to several of the members. It was the first written account of the eclipse that was published in England, as distinguished from accounts telegraphed. It was written by a boy of seventeen, the son of one member and nephew of several others, who went out in the company of a member to visit India and to see the eclipse. The eclipse was seen at sea from the Orient steamship *Orotava* ; and the following account was written by the boy to his father : almost the last thing which he wrote, for he was taken ill a few days later, and died at Constantinople on his way home.

MR. CHARLES F. L. MAUNDER'S ACCOUNT.

THE eventful 22nd January dawned beautifully : except for a few fleecy clouds the sky was perfectly clear. The ship had been slowing down considerably during the last twenty-four hours, and to ensure being on the central line of totality had gone some way out of its ordinary course; in fact, it seemed almost at our disposal,—a private yacht could not have suited us better.

The general excitement was considerably heightened when, on sitting down to breakfast, we found on each plate a bit of smoked glass; and to ensure our getting a good view all the awnings were taken down, so we had nothing to complain of, not even the weather. The sun was shining serenely as ever, and the sea was calm as we could have wished.

At about ten o'clock a clergyman was asked to give a few words of explanation concerning the phenomenon, but he had scarcely commenced when the alarm was given that the eclipse had begun. This took us completely by surprise, for we had been told officially that morning that the eclipse would begin at 11.58; but it was evidently totality that was meant. It was most amusing to see how that meeting broke up; lecturer and audience rushed away; and sure enough, when we looked through our smoked glass there was a part of the sun's circumference obscured.

The progress of the moon was scarcely perceptible; but gradually the sun was blotted out, the temperature fell considerably, and the light began to grow less. About five minutes before totality a most uncanny light prevailed; there was a kind of dull greyish hue over everything, and the general appearance of the sea was like that dead calm which marks the approach of a great storm; and then, as we watched, we could see a silvery light suddenly appear all round the moon: a moment later, and the sun was entirely blotted out.

The sky overhead was dark blue, fading into grey, and finally into lilac near the horizon. The whole effect was not unlike a sunset, only the same colours were visible all round the horizon. A few stars could be seen in the neighbourhood of the eclipsed luminary, but at the darkest it was not necessary to have any artificial light to read or write.

The corona itself was a magnificent sight. All round the moon beams of silvery light shone out, reaching in some places to about twice the moon's diameter. The 2 minutes 9½ seconds of totality was all too short to admire the beautiful sight; and almost before we had taken it all in a beam of light shot out from the other side of the moon, and seemed to chase away the darkness immediately: the sky and sea soon assumed their natural colours, and the eclipse of 1898 was over.

Our position at the time was 12°56′ N. lat., 68°31′ E. long.

CHAPTER VII.

PHOTOGRAPHS OF THE CORONA.

THE work which was most widely taken up by members of the expeditions was that of photographing the corona. From the very first commencement of the photographic art, it was felt that it possessed some great advantages over eye-drawings, particularly as the photographic plate took in the whole of the phenomenon exposed to it so far as this was bright enough to give an impression; and it was, of course, entirely free from the effect of any preconceived ideas. To quote Professor Langley, "The camera has no nerves, and what it sets down we may rely on."

So early as 1851, in the eclipse the track of which passed over Norway and eastern Europe, on July 28th, Dr. Busch took a daguerreotype of the corona at Königsberg. This photograph shows a number of prominences and the inner corona very fairly well to a height of some 8' or 9' from the moon's limb. Mr. J. A. Whipple, eighteen years later, was equally successful with a collodion negative, obtained at Shelby-ville, Kentucky; and the following year Mr. A. Brothers, F.R.A.S., and a member of this Association, obtained a photo-graph at Syracuse of the eclipse of 1870, December 22, which shows not only the inner corona, but the outer to a very great distance from the limb. Indeed, Mr. Brothers' photograph held the record as to coronal extension for an entire eclipse cycle. It was not surpassed in this respect till the well-known eighteen-year period brought a return of the eclipse on 1889, January 1.

1871, December 12, yielded the beautiful negatives taken at Baikul by Mr. Davis, the photographer of Lord Lindsay's (now Earl of Crawford and Balcarres) expedition. These photo-graphs still remain unsurpassed for the beauty and delicacy of detail which they give in the corona.

The corona of 1878, July 29, was successfully photographed at several stations, particularly at Creston, Wyoming; the peculiarity in these photographs being the sharp definition of the edges of the great equatorial wings. These photographs furnish an extreme contrast to those of 1871 in the character

MRS. WALTER MAUNDER'S OBSERVING-HUT AT TALNI.

of the corona which they depict, 1871 being—as remarked in
the preceding chapter—just after the sun-spot maximum, and
1878 being almost dead minimum. The wonderful complexity
of structure, shown by the coronal photographs in the earlier
year, was replaced accordingly by an almost severe simplicity in
the later.

This eclipse of 1878 offered a new problem in coronal
photography. Two of the observers of that eclipse—Professor
Newcomb in Wyoming, and Professor Langley on Pike's Peak,
Colorado—observed the great equatorial wings of the corona
extending eleven millions of miles east and west of the sun,
whilst the photographs do not show them further than one
million. It became a matter of some interest, therefore, to see
if it would not be possible to photograph these extensions ; and
Father Perry, in the eclipse expedition of 1889, December 22
(which cost him his life), included the attempt to photograph
these streamers in his programme.

In the eclipses from 1870 to 1878 the old wet-plate collodion
process was used. In more recent years, gelatine dry-plates
have almost entirely superseded them. From 1882 and onwards
the official expeditions sent out by the British Government
have in every case been provided with one or two photographic
telescopes of 4 inches aperture and about 5 feet focal length,
and a valuable and homogeneous series of negatives have been
accumulated by their means. In 1893 Professor Schaeberle,
from the Lick Observatory, introduced in his station in Chili
the use of a lens of very long focus, giving an image of the
sun over 4 inches in diameter, and obviating the difficulty of
mounting a telescope 40 feet in length equatorially by making
it a fixture, and simply moving the plate. In the same eclipse
the British official expeditions were supplied with "double
cameras." These "double cameras" each carried a pair of
4-inch photographic lenses, and the plates were exposed in the
primary focus of one of these lenses so as to carry on the series
begun in 1882, whilst the other lens was fitted with a Dallmeyer
telephoto lens, so as to give an image of the corona on a scale
about two and a half times as great as the companion negative
in the primary focus. The latter principle was also adopted by
the Astronomer Royal in the Indian eclipse of 1898 to give a
picture on the 4-inch scale with an object glass of 9 inches
aperture and nearly 9 feet focal length.

The eclipse of 1896, August 9, yielded a successful attempt
to photograph the coronal rays at Novaya Zemlya, where
M. Hansky, of the Russian expedition, secured a coronal streamer
some two solar diameters in length with a lens of $2\frac{1}{2}$ inches
aperture and $12\frac{1}{2}$ inches focus.

1896 saw another modification in eclipse instruments, namely
the use of telescopes rigidly mounted and fed by a " cœlostat,"

or mirror, revolving round an axis parallel to the axis of the earth, and rotating at one-half the earth's speed. This method is of great importance where it is necessary to take a number of photographs, and consequently to change the plates with great rapidity, as the shake, to which an equatorially mounted telescope would be liable, is avoided. The plate on p. 81 shows an instrument in Professor Naegamvala's camp used in this manner in conjunction with a cœlostatic mirror.

Practically all these various forms of instruments, beside several other modifications were used at one station or another in the photography of the corona of 1898. The large instruments were, of course, contributed by great public observatories; the members of the Association, having only their own small private resources to draw upon, were necessarily obliged to be content for the most part with apparatus of quite small dimensions. The details of the cameras employed are as follows.

DETAILS OF INSTRUMENTS AND EXPOSURES.

The instrument which I took out with me was an ordinary refractor of 4·1 inch aperture and 60 inches focal length. This I had fitted with a small camera of my own manufacture. I made two exposures; one shortly after the beginning of totality of one second, the other of two seconds. The plates used were Ilford ordinary. Both photographs are fully exposed, and show the polar plumes and the great streamers very distinctly.

FRED BACON.

My camera had a portrait lens, and as used had a focal length of 26 in., its full aperture of $2\frac{3}{8}$ in. being reduced to $1\frac{3}{16}$ in. Four plates were taken; two on Ilford ordinary, backed, and two on Edwards' extra rapid Isochromatic. The exposures varied from one to four seconds, and the plates were developed by pyroammonia. The photographs show only the inner corona.

GERTRUDE BACON.

(Communicated by Mrs. Walter Maunder.)

My lens was a Dallmeyer rapid rectilinear, the back lens of which only was used. As thus used it had an aperture of $2\frac{1}{8}$ in. and a focal length of $32\frac{1}{2}$ in. The camera was home-made, being nothing more than a long light-tight box fitted to carry half-plate slides. Along the top of the camera was fixed a brass tube parallel to the major axis of the camera, having cross hairs fixed at the lens end. A few inches from the other end was placed a little mirror, so that when standing over the camera a person by looking into it could see the cross-hairs up the tube, and also see when the sun was centrally on them.   In the photograph, I am looking into this mirror as I turn the tangent screw.   An ordinary screw held the camera towards the lens end fixed to the stand, while toward the plate end of the camera was fixed a long camera screw, which, acting as a tangent-screw, enabled me to follow the sun and keep it on the cross-hairs. The plates used were Ilford extra rapid red label, half-plate size, thickly backed with Tyler's plate-backing to prevent halation.   Six exposures were given; the first of about ¼ second,

MR. COUSENS AT JEUR.

and the others in order—½, 1, 2, and 4 seconds.   In the middle of the sixth exposure the sun came out.   The developer was the Ilford formula, freely diluted with water, and half as much again bromide of potassium as usual.   The development was prolonged, and the distilled water used, as well as the developer, were both cooled for a day or two previously by placing wet rags around the bottles.   This was very necessary. The shortest exposure brought out too much corona, and half or a third would have been better in order to show the prominences.   The longer exposures bring out the great streamers very fully.

HENRY COUSENS.

My lens was a Dallmeyer stigmatic lens of 1½ in. aperture and 9 in. focal length.   With this I hoped to secure the extensions of the corona, and consequently intended to give prolonged exposures; but I had neither equatorial nor driving-clock. Both these were kindly supplied by the Council of the Royal Astronomical Society, who placed at the disposal of my husband the photographic telescope and equatorial which the late

Mr. Sidney Waters, F.R.A.S., had bequeathed to them. I had seen this instrument at Vadsö, where Mr. Waters, who had designed it for eclipse observation, had taken it in August 1896, and knew it to be convenient and portable. This placed a second camera at my disposal. The lens was a portrait combination of $2\frac{1}{2}$ in. aperture, used in conjunction with a Dallmeyer telephoto-lens, and then having an equivalent focal length of 8 ft. I mounted my stigmatic lens camera instead of the counterpoise of the Waters equatorial, and was enabled to use

THE LATE MR. SIDNEY WATERS, F.R.A.S., AT VADSÖ.

both instruments through the very kind assistance of Captain P. B. Molesworth, R.E., who worked the Waters camera whilst I was employed on the smaller one.

My plan was to take two similar series of six photographs each, the exposures being graduated so as to make the equivalent exposures very nearly a geometrical series of 1 to 1800, the ratio being about $4\frac{1}{2}$. In other words, the actual exposures with each instrument were in the following order:—1, 5, 20, 20, 5, and 1 seconds. The full aperture was not used with the Waters lens, better definition being obtained when the lens was stopped down to $1\frac{3}{8}$ in. The exposures began at the following times from the commencement of totality :—6, 16, 36, 66, 96, and 111 seconds. The last plate in each camera was spoiled by the return of sunlight.

It will be seen at once that this programme involved that many of the photographs would be valuable only as a guide to future work. Nevertheless, one taken with the Waters on an Ilford extra rapid plate, exposure 5 seconds, and three with the stigmatic lens, are exceedingly successful as photographs. The Waters plate shows a remarkable amount of detail for so small an aperture and so great a magnification ; and the three in the 9-in. camera show the coronal streamers farther than they have ever been photographed before.

A thirteenth plate was exposed in the 9-in. camera about forty seconds after the return of sunlight. The plate was a triple-coated Sandell, and the exposure was for 1½ seconds. It shows the inner corona perfectly distinctly, and the whole of the black disc of the moon is clearly defined, whilst the crescent of sunlight is reversed and strongly solarised. It may be noted that this " carraway seed" of solar light was seen distinctly impressed upon the plate before development was commenced, and yet the corona can be traced round till it almost meets the solar crescent., This is the first time that the corona has been unmistakably photographed in the presence of so much sunshine.

The plates used were of three kinds. Two were Hill Norris' dry collodion Eagle plates. These gave practically no result. Six were Ilford red label extra rapid, and were developed with hydrokinone, the development being restrained and prolonged. The best plate taken with the Waters was one of these. Four were Sandell triple-coated, and two of these show the long rays much farther than any of the others. Their development was by hydrokinone and metol.

ANNIE S. D. MAUNDER.

I used a Dallmeyer rapid rectilinear lens of 1 in. aperture and 11 in. focal length, and exposed a single Ilford plate for ½ second, which I developed with " hintokinone. The plate is fully exposed, and shows a dense image of the corona.

A. E. OAKES.

I used a Watson's Alpha camera for 5 by 4 plates with a rapid rectilinear lens ⅞ in. in diameter, and focal length about 5½ in., a stop of $f/16$ being used. I exposed two plates, Ilford special rapid, which I developed with hydrokinone and eikonogen. The exposure of the first was for 3 seconds, the second 25 seconds. The shorter exposure was ample, and that plate is much better than the other.

F. LYS SMITH.

Three photographs were taken at Talni, Berar, India, with an equatorial having a Cooke's 4¼ in. triple, photo-visual object-glass of 71 in. focal length, which was generously lent to me by Mr. Newbegin, F.R.A.S. Ilford ordinary quarter-plates were used, with exposures of 1½, 9, and 30 seconds.

The reproductions, given on p. 109, almost entirely fail to render the mass of complex detail, with its interlacing and overlapping coronal rays, and the solar prominences, as shown on the original negatives. The south-western ray extended to

Mr. D. O. Morris.   Mr. Thwaites.

MR. THWAITES' OBSERVING HUT AT TALNI.

the edge of the plate, and a greater length could have been secured if the plate had been larger.

The corona was of a delicate silvery white hue; and as the darkness was not so great as had been expected, its contrast with the sky was somewhat lessened; but it was an indescribably beautiful and awe-inspiring spectacle, which fully repaid the observers for the long journey from England, and all their trouble and expense.

C. THWAITES.

Exposure, 1½ seconds.

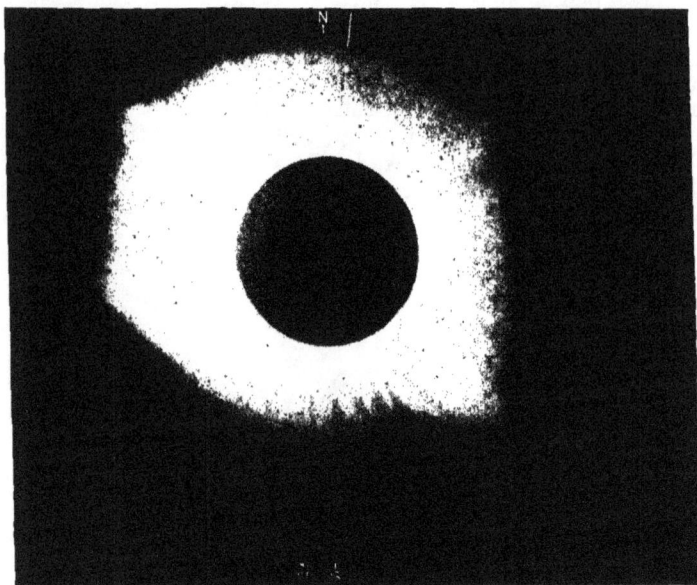

Exposure, 9 seconds.

THE CORONA, 1898, JANUARY 22.
(As photographed at Talni, by Mr. C. Thwaites.)

The following table gives a list of the photographs obtained by members of the expeditions arranged in order of their equivalent exposure; all the photographs being reduced to the standard of a camera with its focal length fifteen times that of its aperture :—

No.	Photographer.	Aperture.	Equivalent Focus.	f/A.	Exposure.	Equivalent Exposure f/15.
		In.	In.		Sec.	Sec.
1, 2	Capt. Molesworth .	1·67	96	57	1	0·067
3	Mr. Cousens . . .	2·1	32·5	15	¼	0·25
4, 5	Capt. Molesworth .	1·67	96	57	5	0·33
6	Miss Bacon	1·2	26	22	1	0·46
7	Mr. Cousens . . .	2·1	32·5	15	½	0·50
8	Miss Bacon	1·2	26	22	2	0·93
9	Mr. Oakes	1·0	11	11	½	0·93
10	Mr. Cousens . . .	2·1	32·5	15	1	1·0
11	Mr. F. Bacon . . .	4·1	60	14⅔	1	1·04
12	Mr. Thwaites . . .	4·25	71	16·7	1½	1·20
13, 14	Capt. Molesworth .	1·67	96	57	20	1·32
15	Miss Bacon	1·2	26	22	3	1·4
16	Miss Bacon	1·2	26	22	4	1·9
17	Mr. Cousens . . .	2·1	32·5	15	2	2·0
18	Mr. F. Bacon . . .	4·1	60	14¾	2	2·2
19	Mr. Lys Smith . . .	0·3	55	16	3	2·7
20, 21	Mr. Cousens . . .	2·1	32·5	15	4	4·0
22, 23	Mrs. Maunder . . .	1·5	9	6	1	6·2
24	Mr. Thwaites . . .	4·25	71	16·7	9	7·25
25	Mr. Lys Smith . . .	0·3	5·5	16	25	22·2
26	Mr. Thwaites . . .	4·25	71	16·7	30	24·2
27, 28	Mrs. Maunder . . .	1·5	9	6	5	31·2
29, 30	Mrs. Maunder . . .	1·5	9	6	20	125·0

Plates Nos. 1 and 2 gave no result, and Nos. 21 and 23 were spoiled by the return of sunlight.

In the above table no allowance has been made for the rapidity of the plates used.

The above photographs were handed for examination to Mr. W. H. Wesley, who has most kindly prepared the diagram which appears as frontispiece to this volume from the photographs, and who has supplied the following report of the appearance of the corona of 1898 from the photographs.

MR. WESLEY'S REPORT ON THE CORONA OF 1898 FROM THE
PHOTOGRAPHS.

THE general aspect of the corona, as shown on the series of
negatives taken in India by the members of the Association,
is that of an irregular four-pointed star. As is indicated by
the outline diagram which forms the frontispiece to the present
volume, the points of the star are formed by the great conical
masses of rays in the north-east, north-west, south-east and
south-west. Consisting of rays curving together, these conical
masses were appropriately named by Mr. Ranyard *synclinal
groups*. These synclinal groups, showing double curvatures on
each side, occupy approximately the same positions as in the
coronas of 1886 and 1896, but present considerable differences
in their structure. Those on the north-east and south-east are
very similar to each other, both being much inclined from the
radial in the equatorial direction. They taper towards their
extremities into somewhat fine, nearly parallel rays, extending
to a distance of 3½ or 4 diameters from the limb. Their broad
bases meet on the limb near the equator, the space between
them being filled by a large tuft of shorter, spreading rays;
this arrangement gives a fish-tail form to the corona on the
eastern side.

The synclinal groups on the west are very different in
character from those on the east, and do not resemble each
other. The north-western synclinal group is ill-defined in its
boundaries. Its principal feature consists of a pair of fine,
perfectly straight, nearly parallel rays, which can be traced
downwards to 6' or 7' from the limb, and extend to a distance of
about 2½ lunar diameters. The most north-easterly of the pair
is radial, and almost exactly 45° from the equator; the other
ray is 4' or 5' distant from it. Fainter rays between and on
each side of the pair follow almost exactly the same direction.
Dense conical masses form the base of these fine rays, but in
one case at least, the mass curves strongly towards the fine ray,
and when very close to it suddenly alters its curvature
into parallelism with it, but does not appear to touch it. This
north-western synclinal group forms the shortest and least conspicuous of the four points of the star to which the corona has
been compared.

A great contrast to this is afforded by the south-western
synclinal group. It is very well marked, very definite in its
boundaries, and appears surrounded by faint envelopes. It is
nearly radial, and its centre is prolonged into a thin, straight
ray, which can be traced on one of the photographs to at least
5½ diameters from the limb. This is not only the longest
extension of the corona of 1898, but is by far the longest that

has ever been photographed. The bases of the synclinal groups on the western side do not reach so near the equator as on the east, and the mass of equatorial rays on the western side is, therefore, the larger, but it is similar in character.

The north and south polar rifts are broad—the northern rift occupying about 40° along the limb, and the southern about 60°. They are pretty symmetrically arranged about the sun's axis, and are filled with rays of the characteristic polar type—straight and radial in the centre of the rift, and on either side becoming more curved towards and inclined to parallelism with the great synclinal groups. The latter, in fact, appear to attract all other rays in their neighbourhood towards themselves.

The corona of 1898 resembles none in its general aspect so completely as that of 1868, as drawn by Capt. Bullock, in Celebes. Indeed, had the north-western and south-western rays in this drawing been interchanged, it would have been in all its main features an excellent representation of the corona of 1898.

Compared with that of 1896, it may be said that the corona of 1898 shows less tendency than might have been expected towards change to the characteristic sun-spot-minimum type. On the east side the depression of the synclinal groups towards the equator approximates to the character presented by that type, but the radial direction of the corresponding groups on the west shows marked divergence from it. At the same time it must be noticed that in 1896 the equatorial regions were of singular interest, showing a very contorted and perturbed appearance, in marked contrast with 1898, when the same regions seemed extremely quiescent. This appears to me a sign of approach to the type of corona generally associated with a period of sun-spot minimum.

I have carefully examined the photographs for traces of the *dark* markings which were so interesting a feature of the corona of 1896. I have failed to find any, except that a few of the rays in the north polar rift appear to be faintly *outlined* on their sides with dark lines (light on the negatives). This is a region, however, in which it is very difficult to distinguish between real dark markings and mere spaces between the bright rays, so I do not like to speak very positively upon the matter.

W. H. WESLEY.

THE CORONA AS SEEN ON PHOTOGRAPHS OF LONG EXPOSURE.[*]

IT will be seen from the table on p. 111 that two photographs (Nos. 29 and 30) taken during the eclipse had a much longer equivalent exposure given to them—more than 2 minutes with $f/15$—than has ever been given before to a coronal negative with plates of corresponding rapidity. These photographs were taken, as already stated, on Sandell Triple-Coated Plates, and were developed according to the formula which follows :—

	grs.
Hydroquinone	20
Metol	2
Sulphite of Soda	130
Bromide of Potassium	6
Citric Acid	3
Carbonate of Soda	90
Caustic Soda	6
Water to 5 oz.	

Two other plates, both taken on Ilford Extra Rapid plates, are interesting to compare with these. These are Mr. Cousens' photograph of 4 seconds exposure (No. 20), and Mrs. Maunder's, No. 27 (equivalent exposure for $f/15$, 31 seconds).

The first two photographs give a presentment of the corona such as has not been yielded by any previous ones. They fail entirely to give anything of the wonderful detail of the inner corona. The scale on which they are taken is too small for that, and their exposure too prolonged. But they give a better idea of the corona as seen by the naked eye than perhaps any other whatsoever. And in so doing they afford a strong and unanticipated confirmation of the truthfulness of a large number of eclipse drawings.

Sketches of the corona have fallen not a little into disrepute of recent years ; partly because of the serious difficulties that interfere with the production of faithful coronal drawings, but principally because the forms ordinarily presented in such drawings have not received support from photographs. The corona as registered by eye and hand, and the corona as registered by the sensitive plate, seemed to be two different structures.

The present photographs give, we think, the explanation of this seeming discrepancy, which resembles in its nature that between the two artists, Mr. and Miss Hall, at the 1874 eclipse, alluded to on p. 88. For if we compare the corona as shown on these photographs with the drawings made by the sketchers at Buxar, we can have no doubt as to the general accuracy of the latter. It is clear that if we had had no photographs

* By Mr. and Mrs. Walter Maunder.

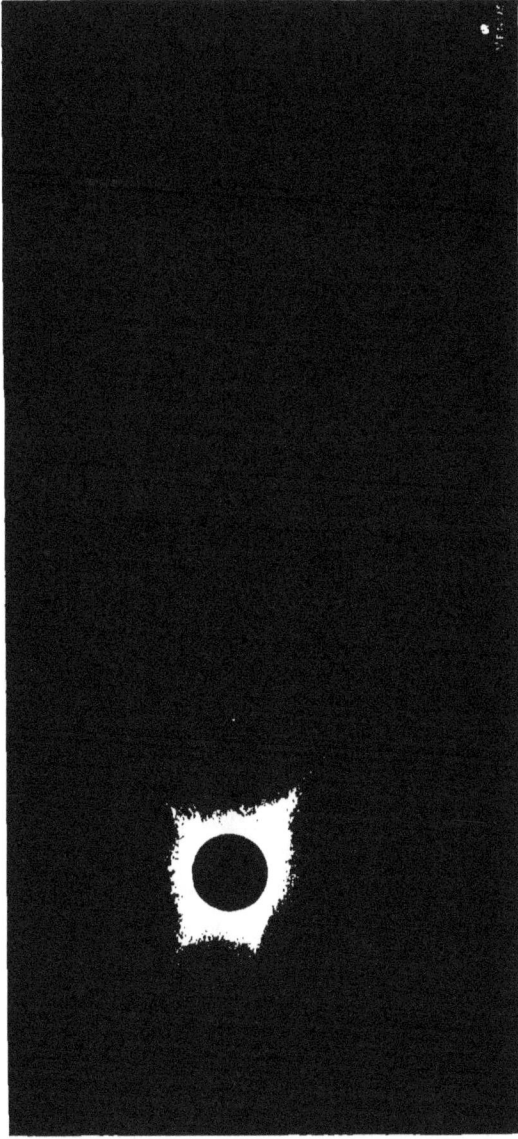

THE CORONAL STREAMERS.

(Drawing by Mr. W. H. Wesley, from photographs by Mrs. Walter Maunder. Reproduced from *Knowledge*, May 1898, by permission of the Publishers.)

whatsoever of this eclipse we should have been justified in placing a large measure of confidence in the features presented to us by the consensus of the Buxar artists. Further, it so closely resembles the well-known drawing by Captain Bullock of the corona of 1868, that we need have not the slightest hesitation in accepting the latter as a faithful representation of the eclipse of that year.

The difference between the drawings and the photographs has no doubt been chiefly due to the very simple fact that the coronal streamers were too faint to be photographed in anything like the time during which the plates had been exposed. Though with adequate exposure a sensitive plate can embrace a wider range of intensity than the eye, yet for such exposures as are alone possible during an eclipse the eye has much the advantage. Never until this eclipse has so long an exposure been given with such rapid plates; and had it been otherwise, it is doubtful whether on any but triple-coated plates the development could or would have been pushed sufficiently far to bring up these faint extensions. The rapidity with which the coronal light faded away was indeed known, but the necessity—if these outer regions were to give an impression—for giving exposures to correspond does not seem to have been sufficiently realised.

The chief features shown by these long-exposure photographs are four long rays.

	N.E.	S.E.	S.W.	N.W.
Position angle from sun's north pole	69°	...	235°	314°
Length on plate No. 20 . .	112′	72′	119′	71′
„ „ „ 27 . .	126	94	162	91
„ „ „ 29 . .	180	123	226	124
Length in lunar radii on plate No. 29	11·0	7·5	13·9	7·6

The position angle of the S.E. ray is not given, as it is very nearly tangential to the sun's limb, whilst the other three rays are nearly radial. The lengths given for the rays are of course their apparent lengths; their real lengths are probably considerably greater, for we do not know in what plane they lie, nor how far their apparent lengths have been diminished by foreshortening; the values given above therefore are a minimum.

The rays in appearance are straight, narrow, and rod-like up to the limits given in the above table. But their bases are of an altogether different form. Each one rises from one of those "synclinal structures" to which Mr. Ranyard called attention in his great eclipse volume (*Memoirs R.A.S.*, vol. xli.). Only four of these structures were seen in this eclipse, and in each case we now see from these photographs that they terminate in one of these rod-like rays. The bending towards each other

of these synclinal curves is therefore not apparent only, as being
due to some effect of perspective, nor accidental, but is of the
very nature of their structure. And we may reasonably infer
from this present evidence that in other eclipses we should
have found—had photographs of sufficient exposure been taken
—that each of the synclinal groups then observed would have
found their completion in similar long rod-like rays.

Turning, for example, to the eclipse of 1871, December 12,
represented by the beautiful photographs taken by Mr. Davis,
at Baikul, and by Mr. Hennessey and Captain Waterhouse at
Dodabetta, we see a number of these synclinal groups distributed
indifferently round the sun, except just at the poles; and in
one of these groups—that at 140° position angle—we notice
the beautiful double curvature which is perhaps the most truly
characteristic of all coronal forms, and which appears again so

COMPOSITE SKETCH FROM THE PHOTOGRAPHS OF THE TOTAL SOLAR ECLIPSE,
1878, JULY 29.

unmistakably in the two groups on the east in the 1898 eclipse.
Mr. Brothers' photograph, taken at Syracuse the preceding year,
shows less distinctly, but to a greater distance from the sun,
a large number of synclinal structures distributed with the
same irregularity as in 1871. We have but to suppose these
terminating in straight rays to understand that to the naked
eye the corona of these two years must have had the strikingly
stellate appearance typical of the solar maximum which so
many of the drawings ascribe to it.

Turning to the great eclipse of 1878, at the sun-spot
minimum, we have an altogether different condition of affairs.
Here we have three synclinal groups, and three only, and these
lie very near the solar equator. The two on the west—like the
two on the east in 1898—form together a fishtail structure,
and the more southerly seems to be projected by the effect of
perspective upon its companion.

Comparing the photographs taken at Creston, Wyoming, with Prof. Newcomb's drawing (*see* p. 164) to which allusion has already been made, when, by the use of a black disk to screen his eyes from the glare of the inner corona, he was able to trace its extent in two great equatorial wings to a distance of twelve diameters from the centre of the sun, it would seem possible that the two great wings which he has described were not wings indeed, but two pairs of straight rays forming the completion of the synclinal groups east and west of the sun. However this may be, the resemblance is certainly great between

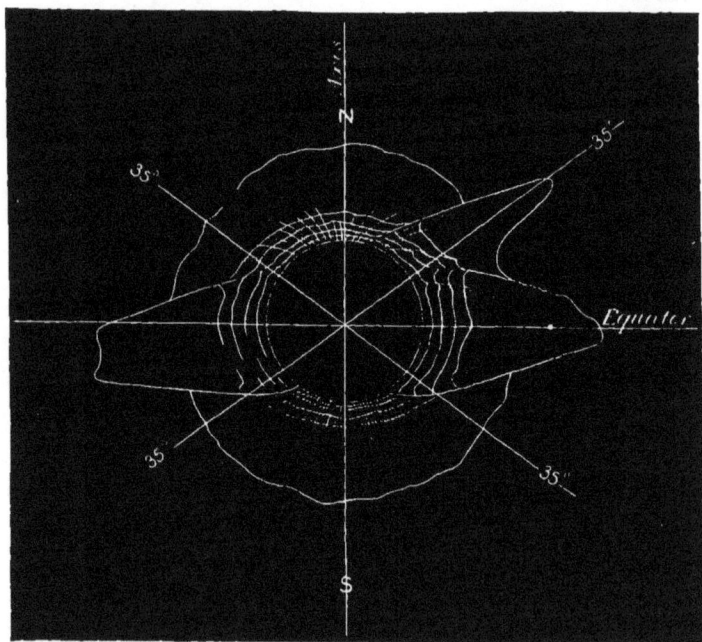

CONTOUR LINES FROM THE PHOTOGRAPHS, 1878, JULY 29.

the double ray in the N.W. in 1878, and the triple straight ray in the same neighbourhood in 1898. In any case we may take it that the essential difference between a maximum and a minimum corona is that at maximum the synclinal groups are numerous and distributed indifferently round the sun, whilst at minimum they are few and are restricted to the neighbourhood of the equator.

The synclinal groups are far from being the only type of coronal form. The 1898 eclipse brought into very beautiful relief those plume-like forms which, because they are seen most easily and distinctly in the neighbourhood of the pole, are generally called the "polar rays." But in truth a careful examination of such photographs as those of Mr. Thwaites and Mr. Fred Bacon, already referred to, shows clearly that similar "plumes" may be seen also in every region of the limb. There can, we think, be little mistake as to the nature of these. On two if not three occasions chance has afforded a very palpable hint as to their nature. In 1882 a small comet was seen near the sun during the eclipse of that year, and it so resembled the coronal "plumes" that, though it was quite detached from the sun, and showed a distinct cometary nucleus, more than one distinguished astronomer was inclined to believe it simply a detached fragment of the corona. Again, in the eclipse of 1893, a comet was actually seen projected upon the corona, from which it would have been almost impossible to distinguish it but for the fortunate circumstance that the interval in time between the taking of the photographs in Chili, Brazil, and West Africa was sufficient for the motion of the comet to render its character unmistakable. Mr. Ranyard also claimed, though on less convincing evidence, that a comet formed part of the coronal forms in 1871. Whether this last was a real instance or not, we may look upon the eclipse comets of 1882 and 1893 as we might upon detached leaves fluttering down from a tree, affording us by their separation an easier demonstration as to the character and nature of those left on the parent stem.

We know perfectly the sort of changes through which a comet passes as it approaches the sun : how thin shells of material rise from the nucleus towards the sun, and are then violently repelled. Professor Bredichin has investigated the forms of a great number of comets' tails, and if his results be accepted we may take it that cometary tails may be grouped in three principal classes, in proportion to the strength of the repulsion exercised upon their material by the sun—a repulsion which is the greater, the lighter the material acted upon. Professor Bredichin therefore considers that the three types of tail correspond to hydrogen, to the hydro-carbons, and to the heavy metals.

What we know of the orbits of comets, and of the meteor streams, leads us to conclude that a vast number of these must have their perihelia within the coronal region. There can, therefore, never be any lack of small bodies in the immediate neighbourhood of the sun to act as cometary nuclei, and to give rise to cometary tails, presenting themselves to us under the form of these beautiful coronal plumes. In this way we can understand the curious coronal rifts, so hard to explain on any other ground.

If the great rays stood alone they might naturally be explained in the same way. A tail, long, narrow, and straight, is necessarily the type which is associated with a comet of extremely small perihelion distance. The comets of 1843 and 1880 are conspicuous examples. But this would leave unexplained the synclinal structure from whence the long rays proceed, and of which they form the continuation. The long-exposure photographs, which we are discussing, necessarily give us absolutely no information as to the way in which these synclinal structures are built up. For that we have to turn again to photographs of larger scale and short exposure.

These show us that over the principal prominences, and at some little distance, an arch of coronal matter is formed. This is succeeded by a larger arch outside, and so on for a succession, the outer arches being less definite and complete than the inner ones. Outside all we find the curves defining the boundaries of the synclinal group.

The impression, which the study of these forms has produced upon us, is that the prominences represent centres of strong eruptive action, and that in consequence of such action coronal matter is driven upward from the sun over a very wide area in dome-like forms. These as they rise would seem to meet with some resistance for a time, a resistance most effective near the circumference of the eruptive area, which averages about 40° of a solar great circle in diameter. But near the centre of this area the eruptive force may be strong enough to overcome the resistance—whatever may occasion it—and therefore from the apex of the synclinal structure we find the coronal matter driven outward in a straight line, which probably indicates an immense velocity. It must be noted that this eruptive action is not always radial. One of the long rays in 1898 was tangential, and another was oblique. Now a ray which is truly radial must appear to be such whatever its position, whilst the effect of perspective, may easily cause lines to appear radial which are not nearly so.

In brief, then, it seems to us that the corona may be considered as threefold in character. First there is the lower corona, pretty evenly distributed round the sun, and showing here and there forms which recall on a somewhat larger scale and with much diminished luminosity the typical forms of the prominences. Next—and it is this section of the corona on which our present photographs have thrown light—we have the synclinal groups rising over disturbed regions, and at least in the best developed instances shooting out at the apex into long rod-like rays. Thirdly, we have a great number of little solar comets scattered indifferently round the sun, but most easily seen near the poles and at the sun-spot minimum. All these three together, projected one upon the other on the apparent vault of the sky, make up for us that complicated structure which we call the corona.

There is another point to which we should like to draw attention in connection with these two photographs, not by way of making a definite assertion, but in order to suggest a subject for observation at the next eclipse. In examining these photographs we have repeatedly been convinced that two rays did not stop at the limits ascribed to them in the table given above, but were continued, though with extremest faintness, much further. These two rays were the S.W. and N.E. The former seemed to us on both photographs to be traceable to fourteen diameters from the centre of the sun—not as a straight line, for if we were not mistaken it became broken a little beyond the point to which it could be easily traced, and then separated into two distinct but nearly parallel branches. It recalled to us, though in a most faint and ghostly manner, Professor Barnard's famous photograph of Brookes' comet.

The difficulty in the way of establishing this extension will be understood when it is borne in mind that the diameter of the sun is but $\frac{1}{17}$ inch upon the photograph, and that the extension, if real, is too faint to permit of the slightest magnification. It can only be seen by most careful adjustment of the illumination. When seen it strikes the eye at first, but is soon lost through visual fatigue. No measurements have been possible, but the form and length of the extension appear to be identical on both the photographs, so far as the most careful eye comparison permits us to judge. If this extension is indeed real, it would seem to offer a further indication of the close connection between cometary and coronal phenomena.

A. S. D. MAUNDER.
E. WALTER MAUNDER.

NOTE ON A PHOTOGRAPH OF THE CORONA TAKEN DURING THE PARTIAL PHASE.*

BEFORE leaving England I took some photographs of the sun on triply-coated plates with various exposures. The remarkably small degree in which these photographs were halated determined me to utilise a spare dark slide, which fitted the camera attached to the stigmatic lens, for a photograph of the sun after totality was over. I had no precedent to go by as to either when I should expose or what length of exposure I should give. The experiment was therefore made—to use a peculiarly inapplicable metaphor—in the dark. The last plates of the series mentioned on p. 106, taken by Captain Molesworth and myself, were surprised by the return of sunlight. The edge of the sun had therefore reappeared by some seconds before we had given the word to the "boys" to cover the object-

* By Mrs. Walter Maunder.

A FUNERAL PROCESSION NEAR BUXAR.

MARKET OUTSIDE THE GATE OF LAHORE.

glasses and we had finished closing the shutters of the dark slides. The "eclipse-clock" was stopped immediately totality was over, so that I had no means of noting exactly the time at which this exposure on the partial phase was given. I had a considerable amount of manipulation to go through—to take out one dark slide and to fit in and latch the other, to cross to the front of the instrument and place the cap on the object-glass, to return to the camera end and draw the shutter of the dark slide, and to again go to the front and expose at the object-glass. I estimated the time I took to do this was not less than 2 minutes, and that the length of the exposure was 1½ seconds. In reality I must have performed these operations very much more rapidly than I believed it possible, for the measurement of the arc of sunlight gives a very much shorter interval than two minutes from the end of totality—not much more, in fact, than a quarter that time. The angle subtended by the cusps

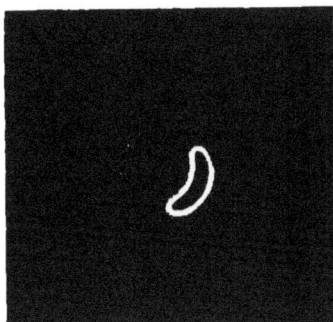

THE INNER CORONA AFTER THE RETURN OF SUNLIGHT.

(Drawing by Mr. W. H. Wesley from a photograph by Mrs. Maunder. Reproduced from *Knowledge*, May 1898, by permission of the Publishers.)

at the centre of the moon is 102° 33', corresponding to a time of 39 seconds after the end of totality. At this time 0·0056 of the sun's disc was exposed to view.

Before the plate was developed, when it was taken out of the dark slide the little solar crescent was plainly to be seen like a carraway seed on the sensitive plate. The intense sunlight had acted chemically on the film, and on development the arc of sunlight became intensely solarised and reversed. There is, however, no serious spreading of the over-exposed arc, and the edge of the moon with its fringe of corona is not fogged out. The greatest height to which the corona is seen is 5 minutes of arc, and marks the position of the N.E. ray. This is shown in the little reproduction attached, and for which I am indebted to the kindness of Mr. Wesley. On the negative the S.E. extension is also plainly visible. The average height of the corona is about 4 minutes of arc.   A. S. D. MAUNDER.

CHAPTER VIII.

BRIGHTNESS OF THE CORONA.

THE corona of 1898 was evidently an unusually bright one, and it is fortunate that several attempts were made in different ways to determine the amount of light which it gave as a whole. These attempts will no doubt be valuable rather as affording suggestions for fuller experiments on similar lines in future eclipses than from the actual conclusions to be drawn from them, yet they will probably give us a more exact idea of the brightness of the late eclipse than we have of any preceding one.

PHOTOGRAPHS OF THE LANDSCAPE.

THE first method consisted of a series of five exposures on the landscape made by Miss Gertrude Bacon with a quarter-plate camera, Lancaster's instantograph, with the lens at the largest stop at $f/10$. The exposures were identical, and were made with a Norden flap-shutter, giving an exposure of one quarter of a second. The exposures were made at intervals of 10 minutes, reckoning from the instant of mid-totality. The first was taken 35 minutes before totality, the second 25 minutes, the third 15 minutes, the fourth 5 minutes before, and the fifth 5 minutes after totality. The most interesting circumstance about the series lies in the fact that the fifth photograph is immensely brighter than the fourth, and, indeed, approximates to the third. The recovery of light therefore seems to have been much more rapid than the decline. That it appeared to be more rapid as judged by the eye has long been known, but this was supposed to be a mere physiological effect. It was not suspected that it was a true objective phenomenon.

The plates employed were Ilford ordinary, and were carefully backed. They were developed simultaneously and for the same time.

From the nature of the photographs it is impossible to reproduce them here. The first is enormously over-exposed, and all detail is lost in one undistinguishable glare. The second and third are nearly what would be counted a correct exposure, especially the second, the third being somewhat lightly exposed. The fourth shows only the sky, against which the objects in the foreground stand up in silhouette. The fifth is distinctly under-exposed, but, as said before, is very much brighter than No. 4.

THE TAJ MAHAL, AGRA.

THE JASMINE TOWER, IN THE FORT, AGRA.

(The Taj is seen in the distance.)

INTEGRATING PHOTOGRAPHS.

Two members of the Association, the Rev. John M. Bacon and Mr. F. Gare arranged for the exposure of photographic plates to the *general* light of the corona during the eclipse, in order to get a measure of its total photographic radiation.

Mr. Bacon exposed an Ilford ordinary plate, quarter-plate size, one half of it on the night of 1897, November 9, at 8^h 15^m p.m., to the light of the full moon at Coldash, near Newbury, and the other half to the light of the corona during the eclipse at Buxar; the exposures being for 80 seconds, and the moon at the same altitude in both instances.

Mr. Gare's plate was a half-plate Lumière orthochromatic, and was placed beneath a sensitometer which he had prepared, divided into thirty squares, one thickness of tissue paper being used in the first square, two in the second, and so on to thirty thicknesses in the thirtieth square.

This plate was exposed by Mr. E. W. Johnson, at Buxar; and Mr. Johnson is seen at his work in the photograph on p. 90. A precisely similar plate was exposed under the same sensitometer by Mr. Gare and Mr. A. H. Johnston, at Dulwich, to the full moon on 1898, April 6, at 10 p.m., the exposure in each case being for 1 minute. The two plates were developed at the same time and under identical conditions by Mr. A. H. Johnston. Six plates from the same batch were exposed under the same sensitometer to the flame of an amyl-acetate lamp at Faraday House, with the kind assistance of Mr. Albert Campbell, by Mr. and Mrs. Walter Maunder; and the density of the deposits on portions of these plates and of that taken by Mr. Bacon were measured by Mr. Walter Maunder with the photometer of the Imperial Dry Plate Company, with the kind permission and assistance of Dr. J. J. and Mrs. Acworth.

The results of the comparison, when reduced to an altitude of 38° for both moon and corona, the moonlight being also reduced in both cases to the mean distances of the earth from both sun and moon and the amyl-acetate lamp to a distance of 1 metre, are as follows for Mr. Bacon's and Mr. Gare's plates respectively.

		Mr. Bacon's plate.	Mr. Gare's plate.
Corona	=	8·9 lamp	4·9 lamp
Moon	=	3·3 „	0·7 „
Corona	=	2·7 moon	7·0 moon.

The difference between the results given by the two plates is no doubt chiefly due to their widely different colour sensitiveness, but partly also to the weather conditions on April 6.

TWILIGHT ILLUMINATION.

IT occurred to two members of the Association, quite independently and in ignorance of what the other was doing, to
attempt to estimate the time at which the general illumination
after sundown corresponded to that during mid-totality.   Mr.
T. W. Backhouse, at Benares, on the day following the eclipse
records: "Watch-time 0^h 40^m.   I judge the light during
totality to have been about equal to that now existing."   A
friend who was with him, Mr. Irwin Sharp, on the following
evening, January 24th, at watch-time 0^h 36^m thought "the
darkness greater now than during totality."   Mr. Backhouse
reported at the same moment that he thought it was not so
dark.

Mr. Walter Maunder, at Nagpur, on January 26th, was
very much struck with the resemblance in the general effect
of the coming on of twilight to the approach of totality,
both as to sky illumination, general light and appearance of
surrounding objects.   He noted at 6^h 15^m Madras time that it
was nearly the same effect as at mid-totality, but on the whole
distinctly brighter ; at 6^h 20^m he noted again that it was distinctly darker than during mid-eclipse.   Taking the mean of
these two times gives the light equal to mid-totality at $30\frac{1}{2}$
minutes after sunset.   Mr. Backhouse's observation of January
23rd gives it at $34\frac{1}{2}$ minutes, the watch being $3\frac{1}{2}$ minutes fast
of Greenwich mean time.   Mr. Backhouse and Mr. Sharp's
observation on January 24th, when combined, gives $30\frac{1}{2}$ minutes.
Mr. Maunder's on January 26th gives $30\frac{1}{2}$ minutes.   The
agreement between the observations made, so entirely without
preparation or knowledge of what to expect, seems to imply
that it was one which can be made with much more precision
than might have been expected, and it will be well worth
while to repeat it on all future occasions of an eclipse in
the tropics.

We may conclude that the general illumination from 30 to 35
minutes after geometrical sunset reproduced very precisely the
illumination at mid-eclipse.

VISUAL IMPRESSIONS AS TO THE BRIGHTNESS OF THE CORONA.

MR. Walter Maunder's note on this point is as follows :—

" In Norway, 1896, there was a slight difficulty in reading
the actual divisions of the little seconds dial of the chronometer, though the seconds hand itself could be readily followed.
In Carriacou in 1886, the writing I had placed on my camera
—a large dark distinct text hand—could just be read.   Now

there was no difficulty about reading anything; and the seconds divisions on the face of the deck watch could be easily seen."

The following sentence in his notebook written at the time may also be worth extracting; though the photographs actually showed longer extensions than the writer had deemed possible under the circumstances:—

"I fear the amount of light in the sky was too great for there to be much hope of getting anything from long exposures or for tracing the streamers for any great distance on the photographs."

Mr. J. Willoughby Meares reported as follows:—

"The brightness of the *inner corona* was far too intense to be compared in any way with the full moon, area for area.

"Even the *streamers* were brighter than the full moon. They appeared to run in straight radial rays over a fainter substratum, and it was not possible to define exactly where they ended, as they grew gradually more like the sky and then merged into it.

"The *general brightness* of the eclipse was certainly as great as the light of two if not three full moons. For (i.) only Venus was visible. (ii.) It was easy to read small print at the normal distance from the eye. (iii.) There was little more difficulty in drawing than in ordinary daylight. Venus was seen 10 minutes before totality and far longer afterwards."

With regard to the visibility of stars during the eclipse, Mr. T. W. Backhouse reports that a friend of his looked specially for stars during totality and saw but three—Mercury, Venus, and Altair; the latter was quite bright, so he has no doubt much fainter ones would have been seen had he had time to look carefully. Mercury was quite plain, he says, like a fourth-magnitude star at night.

Mr. Backhouse's own impression as to the degree of darkness of the sky was that it was far brighter than the full moon.

CHAPTER IX.

MISCELLANEOUS OBSERVATIONS.

BESIDE the observations already detailed, observations of the "shadow bands" have been sent to Mr. Walter Maunder for communication to the Association. These were made at Jeur and Nagpur, and Mr. E. W. Johnson also observed them at Buxar. The positions of these three places together with that of the camp at Talni are given in the following table :—

	Longitude.	Latitude.
Jeur .	75° 10′ E.	18° 15′ N.
Talni .	78° 12′ E.	20° 44′ N.
Nagpur .	79° 8′ E.	21° 9′ N.
Buxar .	84° 0′ E.	25° 33′ N.

OBSERVATIONS OF THE SHADOW-BANDS AT JEUR.

(Communicated by E. Walter Maunder, F.R.A.S.)

HAVING read Professor Naegamvala's pamphlet on "Instructions for Observing the Total Eclipse," we had provided ourselves with a large white table-cloth, which we spread on the grass, and I seated myself beside it so as to be in a favourable position for watching any bands that might appear. The spot our party of ten or twelve friends had chosen for observation was the summit of a slightly rising ground about a mile, or rather more, east of the railway station at Jeur. Some three minutes before totality I noticed for the first time a dim shadow flit across the sheet in the direction from the north-east to the south-west—just the opposite direction to that of the motion of the moon's shadow over the sun's disc. This first shadow-band seemed—for you will remember we had no instruments of measurement, and were all of us merely amateur observers—some four or five inches broad, and its rate of motion quicker than a man can run. Lengthwise the band extended right across the sheet. In three or four seconds after it had passed away another followed, in appearance and with velocity

On the Ganges at Benares.

Jain Temple at Dilwarra.

similar to the first. Others succeeded, but soon the interval of time between successive bands was rapidly reduced, until for a minute or more before totality they followed in extremely rapid succession. I should say that in a second some six or eight travelled across the sheet, and the bands themselves were clearly narrower than those first seen, being now apparently not more than two inches broad, while the interval between every two bands seemed scarcely four inches.

They chased one another with ever increasing speed, all, however, maintaining precisely the same direction, N.E. to S.W., and all parallel each to the others.

As to their intensity, the shadows were never really black,

SKETCH OF THE "SHADOW-BANDS."

only faint and dim, nor did I notice any marked variations of intensity; all, to the naked eye at least, seemed fairly homogeneous. The bands themselves were not clear-cut nor definitely marked, but seemed quivering and tremulous, as though pulsating with life.

After totality had commenced, my gaze was simply riveted on the sun with its wonderful soft silver aureole, and I really can't say whether the shadow-bands still danced on. After totality was over, however, we again noticed them for two or three minutes, during which time their velocity was still great, and the direction was, I believe, but of this I cannot be quite positive, still from N.E. to S.W. But of these post-totality shadow-bands

I should not write anything, for I scarcely paid them any attention. The impression of the glorious sight we had just seen was still upon me, and banished, for the time, thought of all else.

I should perhaps mention that both before and after totality the sky was perfectly clear, and nothing was observable in either the air or the heavens to account for the strange wavering shadows that flitted so noiselessly yet rapidly over the white surface of the sheet.

I can think of nothing to which the phenomenon can be better compared than that which Mrs. Maunder herself mentioned. If the sun's rays reflected from the waves of a calm sea pass through the thick glass of a porthole window, and fall upon the farther wall of a cabin, the faint flickering shadows seen will, in some degree at least, resemble these wonderful and mysterious shadow-bands.

GEORGE P. TAYLOR.

OBSERVATIONS OF THE SHADOW-BANDS AT NAGPUR.

(Communicated by E. Walter Maunder, F.R.A.S.)

FOUR white sheets-were spread on the ground at the top of a small hill.

From two to four minutes before totality the surface of the sheet showed a rippling movement of wavy bands from north-east to south-west, going straight towards the sun; the bands were very faint and appeared to be from $1\frac{1}{2}$ to 2 inches wide; the intervening spaces were a little wider than the bands— perhaps $1\frac{1}{2}$ to $2\frac{1}{2}$ inches. All the bands and all the spaces appeared to be of a constant width. No estimate was made of the rate of movement, but it was rapid, like a fast-flowing tide : the duration may have been a minute or a little more. It seemed impossible to photograph the bands, there was exceedingly little contrast between light and shade.

AGNES E. HENDERSON, M.D.

The "Shadow-bands" were not bands exactly, and were not unlike the shadows cast on the ground by waving branches of trees; they had a rippling kind of movement, and seemed to move up and down as well as across. They reappeared at the close of totality, moving, I think, in the opposite direction.

JOHN LENDRUM.

I wished to observe the sweep of the shadow, the wonderful chiaroscuro of the terrestrial landscapes under the illumination of the chromosphere when the light of the photosphere should be withdrawn, and also the appearance of the corona. Of course we had been reading Sir Norman Lockyer's book, and it had led us to expect a great deal too much, both as to what could be seen and done by untrained observers and as to the striking character of those phenomena that no duffer could help seeing; and our experience in that respect seems to have been shared by other "intelligent but not mathematical" amateurs. The book left one with the impression that sketching the corona was a simple matter, and sketching it presupposes "seeing" it. I only wish Sir Norman Lockyer could be set to "see" a section of pathological liver or kidney (not double-stained) under the microscope and make a sketch of it, too, in a $1\frac{1}{4}$ minutes, and then he would perhaps understand the disappointment which his book prepared for well-meaning but unfortunate amateurs about the corona.

As the light waned a pale violet haze appeared on the horizon chiefly to the westward. Dr. Henderson pointed out to me a green tinge on the land beyond the tank, but I feel sure it was only due to the diminished glare permitting the better appreciation of green bushes and plants among the sere and yellow grass, and the more so as it disappeared instead of deepening towards the climax. We were disappointed in the impressiveness of the coming of the shadow, but in the deeper dusk of totality the light violet haze developed into a bank of heavy dull reddish purple blending above into dull ochreish yellow, which in its turn was lost in the deepened blue of the sky. The sky and its reflection in the tank, the land surface and the trees, were simply darkened by the withdrawal of the daylight.

As regards the corona, I suppose the untrained eye does not know how to "see." I had on spectacles for myopia 2·50 D, so I cannot plead short sight. I saw no details such as Sir Norman Lockyer's questions ask for, but only three long streamers.

As to the shadow-bands, I should rather call them shadow ripples. They reminded me of the figures I have seen while bathing in the Channel Islands when the image or shadow of the ripples on the surface of the clear water dance on the shingly bottom below. Another lady quite independently compared them to the reflection from the surface of the sea on the ceiling of the cabin of a ship. Only the shadow-bands were far less brilliant, smaller, more regular and much less beautiful, but the tremulous, rippling movement was similar. It was a sort of elongated network of shadow, with light spaces some 2 or 3 inches across, and the lengthwise of the network and the direction of the rippling were towards the sun.

EDITH A. HUNTLEY, M.D.

Other observers compared them to the shadow of a barred window falling on the sheet, to the shadows of ripples of water

and to the shadows cast on the wall of a room from rippling
sunlit waters outside, and stated that the shadows were equal
to or narrower than the lighter parts between them.

OBSERVATIONS OF THE SHADOW-BANDS AT BUXAR.

MR. Johnson, who was specially watching for shadow-bands,
first saw them two minutes before totality, their general
direction being W.S.W. and E.N.E. They appeared in batches
of four or five, about 1 inch wide and 3 inches apart. Speed,
say eight to twelve miles an hour. Mr. Cargill independently,
at some distance, recorded the same general direction and speed,
but estimated the bands as 10 inches apart. They were
observed again after totality, but thought not to last so long.

It would seem probable that these faint evasive shadows are
simply due to the effect of air currents at a time when the
sun's light is reduced, immediately before or immediately after
totality, to the thinnest thread. A slight variation in refraction
will then give rise to these ripples of light and shade which will
travel with the wind.

METEOROLOGICAL OBSERVATIONS.

LITTLE serious attention was paid to taking meteorological
observations at the Talni station, but the following may be worth
putting on record. It will be seen from the Talni thermometer
readings that there was a drop in the shade temperature of
10° Fahr. during the eclipse, whilst a fully exposed thermometer
fell 40°. The observations at Buxar were much more complete.

THERMOMETER READINGS AT TALNI, JANUARY 22ND, 1898.

Local Time.	Thermometer Readings.	
h. m.	Full Sun.	Shade.
11 45 a.m.	122°	92°
12 0 ,,	117·6°	91°
12 15 p.m.	113°	91½°
12 30 ,,	109·4°	89°
1 0 ,,	87·8°	88°
1 15 ,,	82·4°	82°
1 20 .,	81·5°	82°

METEOROLOGICAL OBSERVATIONS AT BUXAR.

Hour.	THERMOMETERS.						Cloud.	Aneroid.	WIND. Direction and Force.
	Ground.	Dry.	Wet.	Max.	Min.	Black Bulb.			
Jan. 21.	°	°	°	°	°	°		in.	
10 p.m.	58	56·5	54·5	—	—	—		29·74	
Jan. 22.									
6.30 a.m.	52	50	47·5	—	49	75		29·75	Slight wind, veer-
7.30 ,,	53·5	53	51	—	—	85·2		29·77	ing W. and S.W.,
8.30 ,,	58	59	58	—	—	95			with occasional
9.10 ,,	64	66	59	—	—	97·5			gentle gusts.
9.45 ,,	65·5	67	58·5	—	—	111	None	29·80	Veering S. to N.N.E.
10.15 ,,	67	68·5	61·5	—	—	119		29·79	
10.45 ,,	68	71	63·5	—	—	124·5		29·78	
11.45 ,,	71	73·3	65·2	—	—	128		29·76	
12.30 p.m.	—	76·5	67	77	—	127·5		29·74	
1.0 ,,	71·5	76	67	—	—	109		29·72	
1.15 ,,	72	76	67	—	—	109		29·73	Marked lull.
1.30 ,,	70	75	67	—	—	109		29·72	More northerly.
1.50 ,,	68	71	65	—	—	—		—	

JOHN M. BACON.

GENERAL NOTES MADE AT BUXAR.

AT the Buxar Camp a series of questions were drawn up before the eclipse and supplied to the observers and their friends : these replies were fully discussed after the eclipse, and elicited the following notes :—

THE ZODIACAL LIGHT.—The Zodiacal light was not seen.

THE SHADOW.—The cone of shadow was seen by Colonel Sinclair only, appearing parabolic in outline, rather than circular.

The approach of shadow was generally considered uniform. Colonel Sinclair described it as practically instantaneous (like the drawing up of a shutter).

Mr. Johnson observed one pause in the approach of darkness, followed by a sudden leap into gloom.

THE CRESCENT.—Mr. Hebert, using a 4¼-in. Cooke refractor saw a momentary brush of light from the north end of the crescent, just before totality, about ⅛th diameter long.

No unsteadiness in the crescent was noticed.

BAILY'S BEADS.—Colonel Sinclair observed Baily's Beads through a field-glass for 1 or 2 seconds both before and after totality, and described them as resembling a lady's diamond crescent.

THE CORONA DURING THE PARTIAL PHASE.—After totality several observed the outline of the moon's disc against the corona.

The corona was seen fully 1 second before totality, and afterwards for from 3 to 10 seconds. Finally spikes only were seen in the increasing light.

MISCELLANEOUS.—The moon appeared entirely black throughout.

After totality, the light returned as an intensely brilliant, yellow star and retained this appearance for at least 2 seconds, when it suddenly burst into a blaze.

After the eclipse was over Mr. Meares noticed a hen come down from a tree in which it had gone to roost in a great hurry.

Colonel Sinclair saw the moon on the Friday morning, for 5 minutes from a railway carriage at about 6.5 a.m.

GENERAL NOTES MADE AT DARJEELING.

MR. J. Willoughby Meares forwards the following notes made by friends at Darjeeling :—

The temperature dropped 7° during the eclipse, which was only about ·95 total there, but it appears to have gone nearly as dark as at Buxar, owing to light clouds. (*Miss Curteis.*)

Two friends went from Darjeeling to a mountain about 1200 feet high some 20 miles away, from which Mt. Everest is visible 120 miles off. They saw the shadow strike and leave this mountain, but the mountain remained visible all the time notwithstanding. (*Mr. R. T. Green* and *Mr. S. Shaw.*)

The natives in the Darjeeling "bazaar" were many of them observing the progress of the phase by reflexion from buckets of water. One Bhutia was seen tilting his bucket in every direction in order to bring the water to a more convenient angle! This is vouched for.

ENTRANCE TO THE RESIDENCY, LUCKNOW.

WELL IN WHEELER'S ENTRENCHMENT, CAWNPORE.

CHAPTER X.

PLACES VISITED IN INDIA.

AFTER the eclipse was over, the members of the two expeditions broke up into small parties for the purpose of visiting places of interest in this land—interesting to every traveller, but especially full of interest to Englishmen. In the course of these tours several hundreds of photographs were taken of places, the beauty or historic associations of which attracted attention. A few of these have been kindly supplied by the photographers, who have also added the following brief descriptive notes :—

ENTRANCE TO THE RESIDENCY, LUCKNOW (p. 142).—Since the days of the Mutiny, the famous Lucknow Residency has been converted by a wise Government into a public park, filled with flowers and shrubs, amidst which stand the crumbling ruins of the original buildings, suffered to be touched by no hand save that of time alone. Nevertheless care is taken not to destroy the general features of the old garrison ; and monuments are erected on the site of those houses and defences which have disappeared, so that there is no difficulty in tracing every detail connected with the siege. The photograph shows the ruined archway known as the Bailey Guard Gate. Its mouldering fragments are held together by iron clamps, and its grey walls are swathed in masses of purple Bourganvillia. Upon this, as on every other building, the furious hail of shot and shell has wrought deadly havoc. The Hospital, Dr. Fayrer's house, where Sir Henry Lawrence breathed his last, and the Residency itself, with its range of cellar-like rooms where the women and children passed their weary hours, are the most striking ruins remaining. In the far corner is the cemetery, where lie so many of the gallant defenders, including the hero of the siege, on whose simple tomb are the pathetic words : " Here lies Henry Lawrence, who tried to do his duty."

WELL IN WHEELER'S ENTRENCHMENT, CAWNPORE (p. 142).— This well stands in the centre of the enclosure which marks the site of the entrenchment held so gallantly by the ill-fated European garrison during the terrible days of the Mutiny. The low mud walls, that were then the sole protection of the little force gathered within, have now entirely disappeared, and their place has been taken by a low hedge marking the boundaries; while a few stone pillars show the positions of the frail buildings that stood there at the time of the leaguer. The well alone remains unaltered, and its crumbling woodwork bears the mark of shot and shell. This well provided the only water supply to the entire party of almost a thousand souls through the whole three weeks' siege; and, aware of this, the rebels, with fiendish cruelty, concentrated their heaviest fire upon it. Some of the bravest deeds of that awful time, when every man was a hero and every woman a heroine, were enacted around this spot, ever sacred to English hearts as long as the memory of the pitiful tragedy of Cawnpore shall endure.

THE MAGAZINE AT DELHI.

MAGAZINE AT DELHI.—At the time of the taking of Delhi by the Mutineers, on May 11th, 1857, the magazine, of which the picture represents the sole remaining portion, was held by a party of nine Englishmen under Lieut. Willoughby, described before the Mutiny as "a shy, refined, boyish-looking subaltern," but one of the noblest heroes called forth by that great disaster.

The magazine held a vast quantity of ammunition, much coveted by the rebels, who besieged the building in overwhelming numbers. The little party within held their own with the utmost gallantry for three hours, at the end of which time two of their number were mortally wounded and defence had become hopeless. Willoughby and his men, however, were determined that the powder should not fall into the hands of the enemy; so, when all hope of a rescue was at an end, a train was fired by Conductor Scully, one of the "noble nine," which blew the ammunition, the magazine, hundreds of rebels who stood round, and the brave defenders themselves, into the air together in one awful explosion, in which five of the English lost their lives, and the rest escaped by little short of a miracle.

CASHMERE GATE, DELHI (p. 151).—At the capture of Delhi, on the memorable September 14th, 1857, the Cashmere Gate was the scene of one of the bravest actions of the day. The task of blowing it open, to make an entrance for one of the storming parties, had been allotted to a forlorn hope. A little party of sappers and miners, under Lieutenants Home and Salkeld, made their way, under a deadly fire, across the ruins of a drawbridge that crossed the ditch, each man carrying in his hand a bag of 25 lbs. of gunpowder, which they succeeded in safely depositing under the left arch of the gate. Salkeld, who was to fire the train, was mortally wounded as he bent over it. As he fell he handed the port-fire to a companion, who instantly was shot dead in trying to light the fuse. A third seized the match from his loosening grasp, lighted the train, and in his turn fell, done to death; while the fourth only saved his life by jumping into the ditch as the gate blew open with a violent explosion. A tablet placed beneath the battered wall now bears the names of those who took part in this desperate endeavour.

THE GREAT MOSQUE, DELHI (p. 151).—We were at Delhi on the last Friday in Ramazan, and were allowed to go during the service on the walls and gateway of the great mosque, whence the photograph was taken. The moment chosen is that when the three or four thousand worshippers present, who have all arranged themselves in long lines and have been repeating the prayers after the *imam*, have bowed themselves until their foreheads touch the ground. They do this three or four times just before the close of the service.

THE TAJ MAHAL, AGRA (p. 127).—As is well known, this, the most beautiful and famous building in India, if not in the whole world, was begun A.D. 1630, by Shah Jehan (Great Mogul), as a tomb for his favourite wife, Arjmund Banu, entitled Mumtaz Mahal ("the chosen of the palace"), who died 1629. It

10

occupied twenty thousand workmen for seventeen years in the building, and was erected at a cost of three millions sterling, which it is distressing to think was in great measure never paid. This wondrous tomb has been described as "a dream in white marble," and as "designed by Titans and finished by jewellers." All the work that can be seen is of white marble, much of it inlaid with agates, jasper, bloodstone, etc., all in exquisite taste. It is to-day as fresh and spotless as if but an hour completed ; and standing as it does in beautiful gardens, and seen beneath a cloudless Indian sky, it forms a sight never to be forgotten. By moonlight its fairy-like loveliness is yet more enhanced. Within sleep Shah-Jehan and his much-loved empress. Their tombs are marvels of inlaid jewelling, and are surrounded by a screen of exquisite workmanship. Any noise made within the building will awake a thousand echoes in the great dome above, which answer each other in long-drawn cadence and fade away in softest whispers.

THE JASMINE TOWER, AGRA (p. 127).—The Jasmine Tower (in the Fort of Agra) is a structure of white marble. It was the boudoir of the chief Sultana, the home of Shah Jehan's beautiful Arjmand Banu, who now lies buried in the Taj Mahal, seen in the distance across the river. This tower is built of pure white marble, exquisitely pierced and sculptured, and richly inlaid with semi-precious stones.

THE GHATS, BENARES (p. 30).—The photograph was taken at Benares, on the Ganges, the Sacred River of India. It is a view of the Ghats or Quays (literally stairs). From the river the long range of temples and shrines is very fine. There are forty-seven Ghats, most of them with temples or palaces. The photograph shows some of the roofed river boats ; and on the Ghats may be seen the large umbrellas under which squat the Fakirs or Holy Brahmins, who collect money from those who come down to worship or bathe in the Holy River.

At the time of an eclipse of the sun these Fakirs reap a rich harvest from the millions of pilgrims who journey from all parts of India to the Holy City, so that they may be present at Benares and bathe in the Ganges during the eclipse, and emerge from the water cleansed from all their sins.

One of the Ghats by the river is used by the Hindus for burning their dead, after which their ashes are thrown into the Holy River. Immediately to the right of the trees is the palace of the Rajah of Nagpur ; above it shows the tip of one of the minarets of Aurangzib's Mosque ; and to the right, just beyond the Hindu Temple, is the Ram Temple, the favourite shrine of

PILGRIMS BATHING AT ALLAHABAD.

ON THE GANGES AT BENARES (p. 133).—This photograph gives a closer view of one of the temples at Benares on the river bank. It is a Hindu temple, with one of the antique spirelike roofs of many buttresses arranged in pyramidal form.

FAKIRS (p. 159).—The Fakirs, a group of whom is shown in this photograph, are religious fanatics who wander all over India, having no possessions, but living on charity. They wear very little clothing, but usually smear their bodies over with mud or ashes ; and their hair, which is frequently a yard or more in length, as well as their beards, are clotted with mud and dirt ; the figure on the left is an illustration of this.

THE SACRED BULL AT BENARES (p. 159).—This photograph is taken in the very heart of Benares, where carriage driving is impossible, so crowded together are the shrines and buildings. The sacred cows, each with a bell round its neck to warn the passers by, lurched along lazily, pushing us aside in the narrow passages. The photograph shows in a comparatively large square, a covered structure with a huge marble bull painted red in front of it. This structure is the temple of the " Certainty of Salvation." Here the worshipper at Benares, after the round of special shrines he has come to visit, drinks and receives assurance of success in his task ; he then makes an offering to the priests, hangs a garland of flowers round the head of the famous bull sacred to Mahadwa, and wends his way homewards, happy, for he has drunk of the water which still is one of the favoured homes of Siva.

A MARKET OUTSIDE ONE OF THE GATES OF LAHORE (p. 123). —Beyond the gate can be seen a glimpse of one of the most picturesque of the very narrow streets of the city.

JAIN TEMPLE AT DILWARRA (p. 133).—This temple stands on Mount Aboo, and was built about A.D. 1032, by a merchant, Vimala Sah. Dedicated to Parswanatha, a Jain saint. The interior is of elaborately carved white marble brought probably 300 miles, and carried up 4000 feet from the plain.

BATHING SCENE (p. 148).—The accompanying view was taken at Allahabad on the second morning after the eclipse, during the Mela or religious fair which is held there in January at the confluence of the Ganges and Jumna, a specially sacred spot. It is visited by hundreds of thousands of pilgrims for the purpose of bathing in the river and worshipping it, the most auspicious times being at full moon and during eclipses. Throughout India bathing is an important religious ceremony.

CHAPTER XI.

SUMMARY OF RESULTS AND SUGGESTIONS FOR FUTURE WORK.

IT will be seen from what has gone before that the two expeditions organised by the Association to observe the late eclipse met with a most gratifying degree of success, and this in spite of not a few serious difficulties under which they laboured. It had been found impossible to arrange and practise a comprehensive programme of combined work before starting; the plans drawn up for a camp at Masur were demolished on the very eve of leaving England by the spread of the plague in the district of Satara; the *Egypt* party had the very briefest time at their disposal after arriving at their station for getting ready and for drill; whilst the *Ballaarat* party at Talni were seriously short-handed, and had to omit several items of their programme entirely. All these difficulties notwithstanding, the results obtained are full of interest and value, and afford the greatest encouragement for future expeditions, as the following condensed summary will show :—

SUMMARY OF RESULTS.

1. The spectrum of the "Flash" has been successfully photographed on five plates.

2. A new region of the "Flash" spectrum has been explored and mapped.

3. The bright-line spectrum of hydrogen has been followed in the chromosphere to the 30th point of the series, and the wave-lengths shown to agree closely with Balmer's formula.

4. The correction to the received wave-length of the "coronium" line detected by Mr. Fowler has been confirmed.

5. The distribution of "coronium" in the corona has been examined.

6. Several drawings of the corona as seen by the unaided eye have been made.

7. The variation of the form of the corona with the sun-spot cycle has been further illustrated.

8. Twenty-eight photographs of the corona have been taken.

THE CASHMERE GATE, DELHI.

THE GREAT MOSQUE AT DELHI.

9. These include several showing the coronal rays to a greater distance from the sun than they have ever been photographed before.

10. And cast fresh light upon their structure.

11. Beside restoring the credit of many old drawings of eclipses, which had been supposed untrustworthy because of their want of resemblance to the photographs.

12. The wide range of exposures employed furnishes a most valuable guide as to the exposures to be employed in future eclipses.

13. The corona has been definitely photographed during the partial phase.

14. In several different and independent ways a measure has been obtained of the total illumination given by the corona.

The chief suggestions as to future work arising from these results appear to group themselves under the following principal heads :—

Drawings of the corona.

Spectroscopic observations.

The proper exposure to be given to photographs of the corona.

Photographs during the partial phase.

Determinations of the brightness of the corona.

HINTS FOR MAKING DRAWINGS OF THE CORONA.*

1. The party should consist of at least five persons ; four to sketch details of single quadrants, and one (the leader) to sketch rapidly the general features of the corona. The leader will then be able to correct and supplement the work of the quadrant sketchers, when producing the combined sketch. This must be done on the same day, immediately after the eclipse, and in consultation with the whole party ; and all the drawings should be photographed as soon as completed. The drawings themselves, even if faulty, should not be touched after completion.

2. The party should practise together beforehand, each one sketching his own proper quadrant from a corona-drawing suspended at the angular height of the sun. The time of exposure of the drawing should be slightly less than the known duration of the eclipse. By rotating the drawing it may be made to serve for four sketches. The drawing must be well illuminated and clear; but must not be large. The sun and moon are small objects to the eye, and a large drawing would not give useful practice.

3. Experience shows that Mr. Green's suggestion as to

* By H. Keatley Moore.

materials, white chalk on purplish-blue paper, is an admirable one. For practising, brown paper serves very well, if blue paper is scarce.

4. It is important always to practise on the same scale as the final sketch is to be made. It has been found that a half-crown (1¼ inch in diameter) is a very convenient size for the black body of the moon, and this may be always at hand. A circle, being drawn round the half-crown, is bisected vertically and horizontally, and the diameters are produced across the paper. The sketcher has a plumb-line (a bunch of keys at the end of a string will do), with which he divides his model corona vertically as he looks up at it ; and he guesses at the corresponding horizontal division. His quadrant thus fixed, he proceeds to sketch it. When the time is called the leader redraws all four quadrants in one combined sketch ; and by comparison with the original the habitual faults of the sketchers are detected, and in the course of a few practices will disappear.

5. The position of any planet or high-magnitude star very near the sun at the time of eclipse should be accurately ascertained, and its distance measured in terms of the moon's diameter (taken as half a degree), as these facts when made familiar to the whole party will check the supposed direction and extent of any long streamers of the corona.

6. On eclipse day the sketchers should avoid fatiguing their eyes by too much observation of the preceding partial eclipse, and should rest the eyes for the last five minutes, before totality, absolutely. It would be well to close them for the last minute, and open them by a signal at totality. Attention should be paid to the extreme extent and to the colour of the corona at the moment of beginning to draw, when the eye is at its freshest, and consequently is better able to observe these points than after gazing at the very bright inner parts of the corona.

H. KEATLEY MOORE.

SPECTROSCOPIC RESEARCH AT FUTURE ECLIPSES.[*]

AMONG the numerous questions which present themselves for solution at future eclipses perhaps the most vital is that relating to the distribution of the gases in the Flash spectrum layer. Although the entire thickness of the stratum subtends an angle of but one or two seconds of arc, and consequently appears to us at the moments of second and third contact as an extremely fine thread of light, yet it will have to be examined in much greater detail than heretofore if an advance is to be made in our knowledge of the relation which its bright-line spectrum bears to the dark-line Fraünhofer spectrum. The

[*] By J. Evershed, F.R.A.S.

lower depths in which the photospheric clouds are suspended must be separated from the higher levels, to determine the order of succession of the various constituent gases in passing outward from the photosphere through the 800 miles or so of incandescent gases.

Thus it will be of the greatest interest to learn whether in the lowest depths, where the pressure and temperature are greatest, the emission spectrum becomes more nearly the counterpart of the Fraünhofer spectrum. Or whether, on the other hand, the dissociating effects of higher temperatures in these regions give rise to a simpler spectrum differing materially from the dark-line spectrum, as well as from the emission spectrum of the higher regions.

In the photographs recently obtained the bright lines of the Flash represent the integration of the entire 800 miles of depth, and it is not easy to discriminate between high levels and low levels, which may, and probably do, differ very considerably, seeing that the range of temperature and pressure through this depth is likely to be very great.

At first sight it would seem almost hopeless, with instruments of any reasonable dimensions, to perform this detailed analysis of an object subtending so minute an angle; particularly when it is remembered that under the ordinary conditions of an eclipse the advancing edge of the moon traverses the entire depth of the layer in two or three seconds of time, and only a fraction of a second would be available for photographing the lowest strata.

I think nevertheless that it will be possible at future eclipses to get fresh evidence bearing on this question, which should show, at any rate, which way the tendency lies with regard to the lowest strata.

If observing stations were selected, not on the central line of the eclipse, but only a few miles within the north or south limits of totality, it is probable that some important results would be secured. A calculation of the conditions which would obtain at the eclipse of 1900, May 28th, at a station near the limits of the shadow zone, indicates that although totality itself would be a matter of some twenty or thirty seconds only, the duration of the flash spectrum would be many times prolonged, and the covering up and uncovering by the moon would be a very much more deliberate process.

At a station situated so far from the central line that the duration of totality was reduced to one-third the value it would have on the central line, it would probably be possible to observe the Flash spectrum during the whole time of totality—that is, for about thirty seconds at stations in Spain and Portugal. Under these conditions the bright crescent giving the Flash spectrum would be seen to rapidly shift round the limb from second to third contact, these two points being separated by an angle of 39 degrees only. The lowest strata of the layer would of course only be revealed at the moments of the contacts, while at mid-eclipse only the highest limits would be seen. But the comparatively slow rate at which the excessively thin flash spectrum layer would be occulted by the moon and then again uncovered, would evidently afford an excellent opportunity for obtaining a long series of photographs of the spectrum at the various stages; those taken at mid-eclipse giving high-level spectra only, whilst photographs taken fifteen seconds earlier and later would give high- and low-level spectra combined.

But even under these favourable conditions it will still be difficult to separate the high- and low-level spectra unless instruments of great focal length and considerable aperture are employed. The radiation coming from a stratum at the base of the layer, say within 100 miles of the photosphere, may be very intense, yet it will be difficult to photograph on account of the extreme fineness of the spectrum lines; and small apertures would probably entirely fail to get anything but the spectra of the more extensively diffused gases.

At all stations in the Spanish peninsula where the duration is one-third that on the central line, it will happen also that one of the contacts occurs at, or very near, one of the poles of the sun. If north of the central line, third contact will be at the north pole; and if on the southern border of the shadow track, second contact is at the south pole. In either case, therefore, the spectrum of the polar regions would be obtained, and it would be of interest to learn whether the flash spectrum at the poles differs in any way from that at low latitudes hitherto observed; or whether it has the same composition in all parts of the sphere. It would be important from this point of view to carefully compare the spectra obtained at second and third contacts, since these points being separated by 39 degrees would give this range in latitude.

Another important point which future eclipse work will determine is the composition of the flash spectrum with reference to the sun-spot cycle. A comparison which I have made between the flash spectrum photographed by Mr. Shackleton in August 1896, and that obtained by me in January 1898, shows that in the visible part of the spectrum between D and H the two spectra are identical. The former does not extend far enough in the ultra-violet to make the comparison complete, but below H I can find no line on the one which is not also indicated on the other, and *vice versa*; and the relative intensities seem to be the same. But it may well be that photographs taken at opposite phases of the spot cycle would show differences. Judging by the changes which seem to take place in the coronal spectrum, we should expect the flash to be richest in lines at a time of maximum sun spots. It is well known, too, that when spots are numerous the chromosphere gives many more bright lines than are seen at times of minimum activity, the eruptive action which so frequently takes place in the neighbourhood of spots elevating for the time being the low-lying metallic vapours.

A chance photograph of one of these metallic eruptions taken during the moments when the flash spectrum is visible would be of the greatest interest, as it would probably give valuable information as to the constitution of the very lowest strata. Such a favourable chance is perhaps almost too much to hope for, at any rate until an eclipse occurs at a time of very great solar activity. The observation could only be made at a station near the central line of the eclipse, as eruptive prominences are never seen in high solar latitudes.

The experience gained at the recent eclipse with regard to photographic plates is in one respect of great value to the spectroscopist. The advantages attending the use of triple-coated plates was clearly demonstrated by the successful photographs obtained by Mrs. Maunder; and in particular by the coronal photograph taken forty seconds after the sun had reappeared.

Now, in the series of spectrum plates exposed by me, the first was an ordinary single-film isochromatic, whilst the last was a Sandell triple-coated plate : both received two images of the cusp spectrum with exposures of about half a second, the single-film plate just before second contact and the triple just after third contact. Comparing the two results, it is evident that there is a marked difference in favour of the triple-coated plate. The final exposure on the latter was some twenty seconds after the sun had reappeared, the full blaze of the brilliant cusp falling upon the plate; yet no halation effects obscure the delicate fringe of bright lines bordering the spectrum, and I feel confident that longer exposures might have been made without any ill effects, whilst the finer details would have been more strongly impressed. Probably exposures up to two seconds or more on these plates would give excellent results under the same conditions.

In prismatic camera work these out-of-totality photographs are especially valuable, if only on account of the beautifully defined Fraünhofer spectrum which is impressed. This, in effect, forms a most convenient wave-length scale, which greatly facilitates the reduction of the bright-line spectra, and in all future eclipse work with the prismatic camera it is most desirable that these cusp spectra be obtained both before and after totality.

J. EVERSHED.

On the Proper Exposure to be given for Photographs of the Corona.*

One of the most interesting features in the observations of the late eclipse is the enormous variation in the length of the exposures given. To take the two extreme cases, Prof. C. Michie Smith, with an objective of 6 inches aperture and 40 feet focal length, that is with $f/80$, exposed one of his plates for half a second. Mrs. Walter Maunder, with a lens of $1\frac{1}{2}$ inch aperture and 9 inches focus, $i.e.$ $f/6$, exposed two plates for 20 seconds each, If we take it that the equivalent exposure varies inversely as $(\frac{f}{a})^2$, then in the second instance the exposures were more than 7000 times as great as in the first.

Since both these extreme exposures were successful for the respective purposes for which they were designed, it would seem, at first sight, as if any exposure between these two extreme limits might serve. And there can be no doubt that a really skilful photographer, having a clear idea of the practical exposure which he had given, and of the special features which such an exposure was well calculated to bring out, might even succeed in producing a negative not without value with any exposure within this wide range, possibly within a range even wider still; provided always, of course, that the sky was really clear, and that there was an absence of anything like overwhelming atmospheric glare.

In the Indian eclipse of 1898, the weather conditions were everywhere so favourable that this condition was fulfilled throughout. The question, therefore, was one of pure photography, and we may ask whether we have any indications as to what are the suitable exposures to be given in order to secure certain definite pictures of the eclipse.

Accepting the principle that with different instruments the duration of exposure should vary inversely as $(f/a)^2$, in order to produce the same result, we have before us three series of negatives that may give some definite information. The first we will call the Waters series, as it was taken with the telephotographic camera bequeathed by the late Mr. Sidney Waters to the Royal Astronomical Society. Here the equivalent f/a was 57, and the exposures given were 1, 5, and 20 seconds. The second series was intended as a continuation of this, and was taken by Mrs. Walter Maunder with the Dallmeyer stigmatic lens of $f/6$, the exposures being the same as in the Waters series. This we will call the Dallmeyer series. The third is a set of five plates taken by Mr. Henry Cousens, which we will call the Cousens series. The f/a in this series was 15, and the exposures were $\frac{1}{4}$, $\frac{1}{2}$, 1, 2, and 4 seconds. The value of a given exposure in the Waters series was double that of the

* By E. Walter Maunder, F.R.A.S.

A GROUP OF FAKIRS.

THE SACRED BULL AT BENARES.

same exposure given by Prof. Michie Smith in his great 40-foot telescope, whilst an exposure with the Cousens or with the Dallmeyer instruments were equivalent respectively to 15 or 90 times that exposure with the Waters, and 30 or 180 times that given by Dr. Michie Smith.

Now as to results. The $\frac{1}{4}$-second exposure given with the great 40-foot telescope " shows the very beginning of the eclipse with a range of small prominences round nearly half the sun's limb." The Waters 1-second plate, which was therefore four times the one just mentioned, shows not merely the prominences, but a distinct ring of corona, some two or three minutes in height. We may take it, then, that if the object be to secure the prominences alone with little or nothing of the corona, it will not be safe to exceed greatly the exposure of $\frac{1}{4}$ second with $f/80$, or $\frac{1}{80}$ second with $f/15$; the plate being supposed of maximum sensitiveness, like the " Wratten and Wainwright drop shutter" employed by Dr. Michie Smith. This exposure will be sufficient, and anything much longer will be undesirable.

The next plate in order of light efficiency is the first of the Cousens series, and the estimated exposure was $\frac{1}{4}$ second, equivalent to fifteen times that of the $\frac{1}{4}$-second exposure with the Madras 40-foot lens, which we will take as our unit. Here the exposure of the lower corona, up to 3' of arc from the limb, is full. The prominences are seen, but are much over-exposed, and the brightest begins to eat into the limb of the moon. The fainter parts of the corona are also coming out distinctly, and the roots of the three great coronal rays can be traced to a full radius from the limb.

The 5-second Waters, which corresponds to 20 units, and therefore is $\frac{1}{3}$ as efficient again as the Cousens first plate, shows a very considerable development of the corona. The next two Cousens plates and the third Waters have the values of 30, 60 and 80 respectively, but do not greatly extend the dimensions of the corona beyond that of the 5-second Waters. Exposures, therefore, of from 20 to 30 units, or from $\frac{1}{3}$ to $\frac{1}{2}$ of a second with $f/15$, are sufficient to bring up satisfactorily nearly the whole of the corona as ordinarily shown on photographs. The gain in extent by increasing these exposures, with $f/15$, to 1 second and $1\frac{1}{2}$ second, does not appear to be of any important amount, whilst the risk of entirely losing the detail in the lower corona by over-exposure increases rapidly with the prolongation of the time.

The five splendid photographs obtained by Mr. Thwaites and Mr. Fred Bacon fully confirm these conclusions. Mr. Thwaites employed an o.g. of 4·25 inches aperture and 71 inches focus ; Mr. Bacon had a similar aperture but a focal length of 60 inches. The f/a was therefore 16·7 in the first case and $14\frac{2}{3}$ in the

11

second. Mr. Thwaites' exposures were $1\frac{1}{2}$, 9, and 30 seconds, corresponding approximately to 1·2, 7·25 and 24·2 seconds with $f/15$. Mr. Bacon's exposures were 1 second and 2 seconds, his f/a being almost 15. But both observers used Ilford ordinary plates, not extra-rapid. The exposures are therefore equivalent only to one-fourth or one-fifth of what an ultra-sensitive plate would have given. Mr. Thwaites' second plate is very fully exposed, one of the coronal streamers at least reaching the edge of the plate; and though it shows a great wealth of detail, yet over-exposure is beginning to show itself near the limb. Mr. Bacon's two plates are both amply exposed, whilst Mr. Thwaites' third plate is evidently not at all the equal of his second.

It may be remarked that the two observers were well advised to use slow plates; for where there is no urgent need to cut the exposures very short, there can be no doubt that it is the wiser course to use a plate of normal sensitiveness, rather than an extra-rapid, and to lengthen the exposure in proportion. The slower plate is easier to manage and safer to handle; in other words, a restrained and prolonged development can be better employed upon it, and it is less liable to accidental fog; whilst the grain of the deposit is usually finer. Certainly the five plates in question leave nothing to be desired as to detail and beauty.

The fourth and fifth of the Cousens series show a very real development of the coronal rays: so far as we are aware, the greatest development shown on any photographs obtained during the eclipse, except the three which we secured with the Dallmeyer. These two Cousens plates had exposures of 2 and 4 seconds respectively, equivalent to 120 and 240 units. The first of the Dallmeyer series would have been 360 units, but it is spoiled by bad shake. The second and third had exposures 1800 and 7200 on the same scale, or 31 seconds and 125 seconds with $f/15$. The advantage in extension of the second Dallmeyer over the fifth Cousens is small, when it is borne in mind that the equivalent exposure is $7\frac{1}{2}$ times as long. The two plates representing the third stage in the Dallmeyer series carry the rays very considerably further, but their exposure was 30 times that of the fifth Cousens, and they were taken on the Sandell triple-coated plates.

It is precisely in these extreme exposures that the value of the Sandell plate becomes apparent. The problem of photographing the coronal extensions has three elements. The first, the extreme faintness of these rays, can, of course, be overcome by increasing the equivalent exposure in each or all of three different ways, that is by increasing the actual duration of the exposure, by increasing the ratio of aperture to focal length, or by increasing the sensitiveness of the plate. The second element is the presence in close contiguity to these faint rays of an exceedingly brilliant light. To increase the equivalent exposure, therefore, and to do nothing else, is simply to secure the fogging of the plate before the development can bring up the faint lights. The third element is the amount of atmospheric glare present in the immediate neighbourhood of the faint extensions. With a large amount of such glare, or if it diminishes in brightness with distance from the sun more slowly than does the corona, it is clear that a limit will soon be reached beyond which these rays cannot be photographed. Otherwise the problem is to give an exposure sufficient to show the ray, but not long enough to bring up the general glare, and to use plates so prepared that the intense action set up by the inner corona shall not spread far enough to interfere with the outer.

The evidence before us does not enable us to decide whether the limit of exposure in a fine eclipse has yet been reached. It would certainly appear to be a duty to attempt on the next occasion to increase the exposures given this time as far as possible ; although it is perfectly likely that the result of such an experiment will only be to prove that the limit was nearly reached on the present occasion.

There is, however, some ground for hope that if favoured with skies as clear as we had in India, the present record exposures may be greatly increased with success. The attempt to photograph the corona during the partial phase is photographically but a special case of the same problem. The difficulty was not, indeed, to give a long enough exposure, but to avoid the fogging of the plate from the brilliance of the returning sunlight and from the consequent general illumination of the atmosphere. The success of the attempt was most remarkable, and could only have been obtained on a plate possessing the characteristics of the triple-coated Sandell. When it is borne in mind that the exposure given was equivalent to 9 seconds with $f/15$, or 540 units ; that carefully backed plates exposed during mid-totality for periods corresponding to this are strongly halated, and that, surface for surface, the sun is something like 100,000 times as bright as the corona, it is indeed a remarkable result to find that the coronal ring can be traced round practically to meet the cusps of the returning sun, and that the limb of the moon is seen sharply and distinctly defined black against its bright background. With no other plate that I know of would it have been possible to develop a photograph with any hope of success upon which the image of the sun had come out deep, distinct and black before ever the developer touched it. Mr. Cousens, for example, exposed a sixth plate, carefully backed, which was caught by the first ray of returning sunlight. The flood of sunshine made it impossible to fully develop the plate without entirely fogging

it. The negative is consequently thin, and it shows no more
of the corona than the first of the same series with but a
a sixteenth of the exposure.

The Sandell triple-coated plate appears, then, to offer a
prospect of overcoming the second of the three difficulties in
the way of photographing the coronal rays to which we have
already referred ; and the success of this attempt to photograph
the partial phase, whilst it accentuates the extreme difficulty
of the problem of photographing the corona in full sunshine,

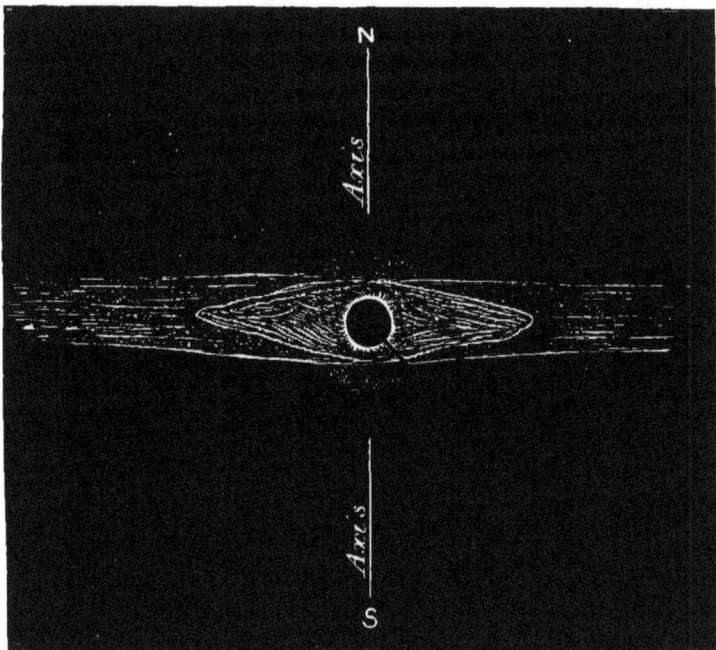

THE CORONA OF 1878, JULY 29.

As drawn by Professor Newcomb, the inner Corona being screened off by a black disc.

does seem to make it probable that during totality, if the sky
be really clear, a plate may be successfully exposed for a much
longer effective duration than has been attempted even on
the present occasion.

Summing up, and taking $f/15$ as a representative instrument,
and supposing that the most rapid plates available are used,
four different classes of exposure appear to be indicated according to the four different regions of which it is desired to obtain
a photograph. For the prominences $\frac{1}{60}$ second to $\frac{1}{15}$ second
may be taken as normal, and for the inner corona $\frac{1}{10}$ second to

$\frac{1}{5}$ second. Prolonging the exposure to $\frac{1}{2}$ or perhaps $\frac{3}{4}$ second, gives a great increase in the area commanded; and 1 second may be taken as marking almost the limit of a really useful exposure for the corona as a whole. If beyond this we wish to secure the long rays, we must be prepared to expose for 100 seconds or upwards, sacrificing, of course, the inner corona for this purpose. It is possible that we may find that exposures of 200 or 300, or even longer, are advantageous. But exposures of above 2 seconds until we get to 50 or 60 or upwards require the most skilful manipulation in development to render them of any service. Otherwise detail in the corona proper is sacrificed without any sensible increase in the representation of the streamers.

One corollary of great importance for amateur photographers results from these considerations. Where f/a is not greater than 15, and the focal length is not greater than 5 feet, the blurring in a stationary camera of a photograph exposed for $\frac{1}{2}$ second is only $\frac{1}{100}$ inch, which is practically imperceptible. A sharp image of the whole of the main body of the corona, inner and outer, can therefore be secured by a fixed camera, and there is no need for equatorial or driving clock, heliostat or cœlostat, with the consequent expense, weight, and trouble of adjustment.

The one and only caution to be borne in mind is that the camera should be firmly and rigidly mounted. This deduction from the results of the 1898 eclipse should lead to a large number of cameras of considerable focal length being used in the eclipse of 1900, resulting in an abundant harvest of fine photographs on a sufficient scale.

It would be well worth while to try and arrange for photographs to be taken on the Appalachian Mountains in the United States, and in Algeria, with instruments of precisely similar make and size. With an interval of $2^h 40^m$ in absolute time between the exposures at the two stations, there would be an opportunity for again examining, as in 1893, if the corona showed change in a short interval of time, whilst the photographs might possibly be combined in pairs in the stereoscope. If this could be done successfully, our knowledge of the true form and structure of the corona would be immensely increased at a glance.

E. WALTER MAUNDER.

PHOTOGRAPHS OF THE PARTIAL PHASE.*

IT would seem worth while to include amongst the subjects for experiment during the next eclipse, photographs of the corona

* By Mr. and Mrs. Walter Maunder.

during the partial phase. There can be no doubt that plates triply-coated, or possibly even with four coats, will be necessary for this work. It seems to us that the chief improvement possible upon our attempt during the late eclipse lies in the direction of very greatly shortening the exposure. There can be no question at present of photographing the long rays or the outer corona in a large amount of sunshine. The only thing that can be hoped for at present is to secure the inner brighter corona. Now for this under favourable conditions $\frac{1}{10}$ or $\frac{1}{5}$ second with $f/15$ is sufficient *during totality*; out of totality a yet shorter exposure will be enough, and it will be well to arrange for a fairly wide range of exposures—say from $\frac{1}{5}$ second down to $\frac{1}{30}$. In all probability it will be found that the duration of the most efficient exposure varies with the amount of the sun's disc uncovered, and that the farther out of totality the photograph is taken, the shorter should be the exposure.

The question of hiding the sun behind an occulting disk may offer some little difficulty; but if a sufficiently good guiding telescope is rigidly attached to the camera, and the disk is the precise size of the image of the sun, it will obviously be a great advantage to use one; but all would depend as to whether the sun could be brought accurately upon the disk at the moment of exposure. This is a crucial point.

The importance of using a disk does not lie only in the fact that it would screen the plate from the direct rays of the sun. It will be seen, on referring again to the little photograph on p. 125, that not only was the sun's image reversed, but it was solarised; and though we may hope to diminish the cause of this on the one hand by immensely shortening the exposure, yet, on the other hand, if we succeed in getting photographs far out of totality, the larger amount of the sun's surface visible would increase it on the other. At all events the presence of this solarised ring is fatal to any chance of showing the corona on the sunlit side of the partial eclipse. On the other hand, the moon itself would cover the whole of the inner corona on the other side except at the cusps and when the partial phase was not far short of totality. This difficulty is not insurmountable, as photographs of the entire disk of the sun have been taken with an ample exposure for securing the corona, and yet without any trace of solarisation at the limb; but it requires very skilful manipulation to avoid it, and the use of an occulting disk would seem to offer the easiest way of securing the desired result.

A. S. D. MAUNDER.
E. WALTER MAUNDER.

DETERMINATIONS OF THE BRIGHTNESS OF THE CORONA.

PHOTOGRAPHS OF THE LANDSCAPES.—No previous experiments of a like nature having been made, it may be considered that the exposures adopted were fortunately chosen for their special purpose of recording the rate of decline of sunlight during the progress of the partial phase and also of giving a comparison between the light of the sky immediately before and that immediately after totality. It should be borne in mind, however, that the darkness during other eclipses seems often to have been more intense, and it would certainly be well to take at least two or three exposures after totality, say at intervals of every two minutes and compare them with similar exposures made at corresponding intervals before totality.

When testing the light of the full moon for comparison with that of the corona it would seem that very careful note should be made of time and place and general conditions. It might be instructive also if such tests were made at more than one lunation. The actual altitude of the place of observation may be of some importance in face of the fact that the intensity of moonlight rapidly increases with elevation on leaving the earth's surface.

JOHN M. BACON.

INTEGRATING PHOTOGRAPHS.—Previous to the eclipse of 1896 the question of obtaining a measurement of the intensity of the light of the corona by means of photography was raised by Mr. Maunder, and experiments with a view to its determination formed part of his programme and that of Mr. Lunt at Vadsö. These experiments, though differing in detail, had for their fundamental idea the exposing a plate behind a graduated screen, and the same principle was made use of in an experiment carried out by Mr. Moore and Mr. Johnson at Benares in 1898.

The experience gained in measuring this and comparison photographs has suggested the possibility that better results may be obtained by discarding the screen and exposing a plate in sections to the light of the corona, giving exposures of varying lengths to the different sections. Trial exposures of similar plates to a standard light could then be made with similar development, and the results compared by means of the density of the deposits on the films.

Whichever method be adopted it should be used at all the stations of the B. A. A. during the eclipse of 1900, under as far as possible identical conditions, and in that case it can hardly fail to give a satisfactory measure of the total light of the corona. The apparatus required is exceedingly simple, so that the experiment would be within the reach of those who do not care to undertake more ambitious work.

F. GARE.

It is on record that two English tourists once discussed the question as to whether it were possible to pay an Irish car-driver so well that, unlike Oliver Twist, he should not "ask for more." The controversy having taken the form of a bet, the more sanguine disputant called a car, rode a quarter of a furlong, and offered the lucky driver a sovereign " in full of all demands." " Ach! shure," says Pat; " but it would be a pity to break it : wouldn't you be giving a sixpence more just to drink your health ! "

We trust that too close a comparison will not be drawn between our Association and this greedy carman, if after a chronicle of so much success it is found that, as if counting what has been done as nothing, it looks forward to organising more expeditions on a larger scale with more ambitious plans, and is expectant of even greater success in the eclipse of 1900.

NEARLY HOME.

A cold day in the Channel.

INDEX.

Printed by Hazell, Watson, & Viney, Ld., London and Aylesbury.